The Creation of Heaven and Earth

Themes in Biblical Narrative

Jewish and Christian Traditions

VOLUME VIII

The Creation of Heaven and Earth

Re-interpretations of Genesis I in the Context of Judaism, Ancient Philosophy, Christianity, and Modern Physics

Edited by

George H. van Kooten

BRILL
LEIDEN · BOSTON
2005

Cover design by TopicA, Antoinette Hanekuyk

This book is printed on acid-free paper.

Library of Congress Cataloging-in-Publication Data

LC Control Number: 2004059618

ISSN 1388-3909
ISBN 90 04 14235 5

Cover illustration: God as the Architect of the Universe, French Moralized Bible (Österreichische Nationalbibliothek, Vienna Codex 2554, Frontispiece), 13th century AD

PRINTED IN THE NETHERLANDS

TABLE OF CONTENTS

PART III

CREATION IN THE JOHANNINE, PETRINE, AND PAULINE LITERATURES

PART IV

CREATION IN THE MIDDLE AGES AND MODERNITY

INTRODUCTION

This volume is about the macro-creation of heaven and earth, as narrated in Genesis 1. It deals both with Genesis 1 itself, and with the interpretations of this narrative in the successive contexts of Judaism, ancient philosophy, Christianity, and modern physics. The articles are the revised versions of papers presented at the Themes in Biblical Narrative (TBN) conference held at Groningen on June 13th-14th, 2003. As such, this theme is the follow-up of the 1999 TBN conference, which was concerned with the micro-creation, the creation of man and woman (Leiden: Brill, 2000; TBN 3).

It was perhaps particularly appropriate to deal with macro-creation (or the lack of it) at a conference in Groningen, in light of Pliny's description of the Northern region, which extends from Frisia to Groningen and Northern-Germany. Pliny, taking the vantage point of the tribes of the Chauci, described what he regarded as the desolation of the Northern people in this region, who had to live on self-constructed earthen mounds which protected them from the tides and flooding of the 'Northern Ocean': 'There twice in each period of a day and a night the Ocean with its vast tide sweeps in a flood over a measureless expanse, covering up Nature's age-long controversy and the region disputed as belonging whether to the land or to the sea' (Pliny, *Natural History* 16.2; Loeb transl. H. Rackham). It was in this former wasteland between sea and earth, in the congenial atmosphere of the new housing of the Faculty of Theology and Religious Studies in the Old Court House, that this conference took place.

By focusing on the macrocosmic aspects of creation, both in the original text of Genesis 1 and its later interpretations, the authors dealt with such diverse issues as the creation of heaven, earth, and light, the Spirit of God hovering over the surface of the water, the role of the separation of the waters in the creation process, and the purposefulness of creation. The papers can be grouped together in four chronological and cultural clusters.

The first cluster, devoted to Genesis 1 and its interpretations in the contexts of the Old Testament, the Ancient Near East and Early Judaism, opens with Ed Noort's analysis of macro-creation in the text of Genesis 1 itself. Noort (Groningen) focuses on the creation of light in this account, and explores the function of the light and darkness antithesis.

Jacques van Ruiten (Groningen), in his turn, studies the reception of Genesis 1 in an oracle of judgement in Jeremiah, in which the prophet announces the reversion of creation back into primordial dark chaos, waste, and void (Jer 4:23–26).

Eibert Tigchelaar (Groningen) comments on the link between the creation of light itself on the first day and the subsequent creation, on the fourth day, of the luminaries, whose function is to serve 'as signs for festivals' (Gen 1:14b). Tigchelaar draws comparisons with the function of luminaries according to texts from the Ancient Near East, particularly the *Enūma eliš*, and goes into the calendar controversies of Early Judaism which affected the way in which Gen 1:14 was read in this period.

Finally, Florentino García Martínez (Groningen & Catholic University of Leuven) gives a separate treatment of creation in the Dead Sea Scrolls and argues that 'creation,' as a abstract concept still absent from the Hebrew Bible, had already reached a particular level of abstraction at Qumran.

The second cluster is concerned with the contexts and contents of the interpretation of Genesis 1 among the Greeks. Jan Bremmer (Groningen) paves the way with an extensive comparative study of Greek mythology, Ancient Near Eastern cosmogonic myths, and Genesis 1. Having discussed the 'canonical' creation myths of Homer and Hesiod, he continues with the alternative versions offered by Euripides and Aristophanes, to conclude with an appendix in which he argues that the beginning of Genesis reveals Persian influences. The outcome of this entire comparative study suggests that the cosmogonic account in Genesis and Hesiod attained a degree of rationalization which is absent from the Near Eastern accounts, and that, whereas 'in the Near East cosmogony was closely associated with ritual, Israel and Greece emancipated themselves in this respect, which may well have enabled the turn towards philosophy that we witness in ancient Greece' (Bremmer, this volume, §3). This turn and the subsequent interaction between Greece and Israel is the topic of the following papers in this cluster.

John Dillon (Trinity College Dublin) examines the way in which Philo of Alexandria participated in the ancient philosophical debate on Plato's *Timaeus* and the consequences of this for his interpretation of Genesis 1. In this, Philo appears to have been stimulated by the Greek

Septuagint translation of Genesis 1, which 'lends itself more readily to philosophical reinterpretation' (Dillon, this volume, §2).

Not only were Jews such as Philo familiar with Greek ideas about creation; the creation account of Genesis was also known to Greek philosophers. The earliest encounters which have survived are the subject of two papers. Robbert van den Berg (University of Leiden) deals with the second century AD Platonist philosopher Numenius and his quotation and interpretation of Gen 1:2—a passage which mentions that 'the Spirit of God was borne upon the waters'. Van den Berg reflects on Numenius' interest in the Jewish Scriptures and offers an explanation of Numenius' interpretation of Genesis 1 against the background of his philosophy.

Teun Tieleman (University of Utrecht), in his turn, reconstructs the context in which the physician-philosopher Galen makes reference to Moses' views on creation in his *On the Usefulness of Parts*. This work (169–176 AD) was intended to demonstrate that the human body had been designed, and to offer an extensive argument, from this design, for God's existence. Tieleman's reconstruction shows that Galen refers to Moses as part of his strategy against Epicurus' atomistic, mechanistic cosmogony, from which the Creator is omitted.

The third cluster offers papers on the reception of Genesis 1 within the New Testament. The articles deal with the Johannine, Petrine and Pauline strands of earliest Christianity respectively. George van Kooten (Groningen) sets out to demonstrate that the 'true light' in the Prologue of John's Gospel refers back to the light of creation in Gen 1:3–4 and is interpreted in terms of the Platonic notion of the 'true, noetic light'.

Edward Adams (King's College London) offers a reconstruction of the polemics between the author of 2 Peter and his opponents. In this polemic, Second Peter understands the creation of Genesis 1 as a creation which was 'created by God's word out of water and through water' (3:5). Interpreting these words against the background of Stoic cosmology, Adam concludes that Second Peter offers a Stoic re-interpretation of the Genesis creation narrative. Adams even interprets the controversy in 2 Peter about creation as a reflection of the wider philosophical dispute about the cosmos between Platonists and Aristotelians, on the one hand, and Stoics, on the other. It serves to show that in their cosmological views, too, the earliest Christians were fully integrated into the Graeco-Roman world.

Boudewijn Dehandschutter (Catholic University of Leuven) highlights another aspect of the earliest Christian reception of Genesis 1. In his paper on the first Pauline Letter to Timothy, Dehandschutter discusses the background, in terms of the history of religions, of the letter's insistence on the fact that 'everything that God has created is good' (4:4).

The fourth and final cluster contains papers on the reception of Genesis in the Middle Ages and Modernity. Willemien Otten (University of Utrecht) traces the development from Augustine through John the Scot Eriugena up to the School of Chartres in the twelfth century, and shows how the Book of Genesis and Plato's *Timaeus* constitute the co-ordinate system in which Nature is studied.

The last paper, by René van Woudenberg (Free University of Amsterdam) brings the discussion of Genesis and creation up to the present day by paying attention to the current debate about design in Nature. As is apparent from Tieleman's paper, this interest in design was already characteristic of ancient philosophers such as Galen. In a reply to Van Woudenberg's paper, John Dillon gives some comments on the modern discussion from the perspective of ancient philosophy. This reply may serve to underscore the continuity between modern and ancient positions. In this sense, it may save us from chronological snobbery.

Finally, I wish to thank the Faculty of Theology and Religious Studies of the University of Groningen and the Groningen Research School for the Study of the Humanities for their moral and financial support, without which this conference would have remained *in nihilo*. I also express my gratitude to Freek van der Steen, assistant series editor for Brill Publishers, and the staff at Brill for taking excellent care that this volume should indeed materialize. Pieter Nanninga, my student assistant, was of great help in compiling the index of ancient texts.

Groningen, July 2004

George H. van Kooten

University of Groningen
Faculty of Theology & Religious Studies
Oude Boteringestraat 38
9712 GK Groningen
The Netherlands
Website: www.rug.nl/theology

LIST OF PRINCIPAL ABBREVIATIONS

AHw	W. von Soden, *Akkadisches Handwörterbuch*, 3 vols (Wiesbaden, 1965–1981)
ANET	J.B. Pritchard (ed.), *Ancient Near Eastern Texts Relating to the Old Testament* (Princeton, NJ 1950; 1955²; 1969³ with supplement)
BHK	Biblia Hebraica Kittel
BHS	Biblia Hebraica Stuttgartensia
CAD	I.J. Gelb *et al.* (eds), *The Assyrian Dictionary of The Oriental Institute of the University of Chicago* (Chicago, 1956–)
DDD	K. van der Toorn, B. Becking, P.W. van der Horst (eds), *Dictionary of Deities and Demons in the Bible* (Leiden, 1995; 1998²)
Diels-Kranz	H. Diels and W. Kranz, *Die Fragmente der Vorsokratiker: Griechisch und deutsch*, 3 vols (Zürich & Berlin, 1964¹¹)
DSSSE	F. García Martínez and E.J.C. Tigchelaar, *The Dead Sea Scrolls: Study Edition* (Leiden, 1997–1998; 2000)
DJD	Discoveries in the Judaean Desert (Oxford)
Ee	Enūma eliš
FGrH	F. Jacoby, *Die Fragmente der griechischen Historiker* (Leiden 1957–1964, second edn)
JBL	Journal of Biblical Literature
JJS	Journal of Jewish Studies
Jub.	Jubilees
KUB	H.G. Güterbock, *Kumarbi: Mythen vom churritischen Kronos aus den hethitischen Fragmenten zusammengestellt, übersetzt und erklärt* (Zürich, 1946)
LXX	Septuagint
MS	Manuscript
MT	Masoretic Text
NRSV	The New Revised Standard Version of the Bible
OCD	S. Hornblower and A. Spawforth, *The Oxford Classical Dictionary* (Oxford & New York, 1996³)
OF	Orphicorum Fragmenta
PAM	Palestine Archaeological Museum
PL	J.-P. Migne, *Patrologiae Cursus, Series Latina* (Paris, 1844–1890)

PMG	D.L. Page, *Poetae Melici Graeci* (Oxford, 1962)
Sir	Sirach
TUAT	O. Kaiser (ed.), *Texte aus der Umwelt des Alten Testaments* (Gütersloh, 1982–2001)

CONTRIBUTORS

EDWARD ADAMS, Lecturer in New Testament Studies, Department of Theology and Religious Studies, King's College London

ROBBERT M. VAN DEN BERG, Post-Doctoral Researcher of the Netherlands Organization for Scientific Research (NWO) & Lecturer in Ancient Philosophy, Department of Classics, University of Leiden

JAN N. BREMMER, Professor of the General History of Religion and Comparative Religious Studies, Faculty of Theology and Religious Studies, University of Groningen

BOUDEWIJN DEHANDSCHUTTER, Professor of Church History and Historical Theology, Faculty of Theology, Catholic University of Leuven

JOHN DILLON, Regius Professor of Greek, School of Classics, Trinity College Dublin

FLORENTINO GARCÍA MARTÍNEZ, Professor of the Religion and Literature of Early Judaism & Director of the Qumran Institute, Faculty of Theology and Religious Studies, University of Groningen, and Research Professor of Theology, Catholic University of Leuven

GEORGE H. VAN KOOTEN, University Lecturer in New Testament and Early Christian Studies, Faculty of Theology and Religious Studies, University of Groningen

ED NOORT, Professor of Ancient Israelite Literature, Old Testament Interpretation, the History of Israelite Religion and Intertestamental Literature, Faculty of Theology and Religious Studies, University of Groningen

WILLEMIEN OTTEN, Professor of the History of the Church, Department of Theology, University of Utrecht

JACQUES T.A.G.M. VAN RUITEN, Senior University Lecturer in Ancient Israelite Literature, Old Testament Interpretation, and Early Jewish Literature, Faculty of Theology and Religious Studies, University of Groningen

TEUN TIELEMAN, University Lecturer in Ancient Philosophy, Department of Philosophy, University of Utrecht

Eibert J.C. Tigchelaar, Fellow of the Netherlands Organization for Scientific Research (NWO) at the Qumran Institute, Faculty of Theology and Religious Studies, University of Groningen

René van Woudenberg, Professor of Epistemology and Ontology, Faculty of Philosophy, Free University of Amsterdam

PART I

CREATION IN GENESIS, JEREMIAH, THE ANCIENT NEAR EAST, AND EARLY JUDAISM

THE CREATION OF LIGHT IN GENESIS 1:1–5: REMARKS ON THE FUNCTION OF LIGHT AND DARKNESS IN THE OPENING VERSES OF THE HEBREW BIBLE

Ed Noort

Introduction

The Priestly account of the creation of the world in Gen 1:1–2:4a by the word of Elohim is one of the best known and most studied texts of the Hebrew Bible. Nevertheless, there are many things we still do not know. In his commentary, Horst Seebass states concisely:

'Am berühmtesten ist der gewaltige Schöpfungsentwurf 1,1–2,3 für den eine Formbestimmung nicht gelungen ist'.[1] This means that a convincing function for the text in relation to its *Sitz im Leben* has not yet been demonstrated.

Was it recited at an Israelite, or rather Judaean, New Year Festival as *Enuma elish* was?[2] The strong, even monotonous, repetition of certain elements, and its sophisticated character as a 'litany', would favour such a rhetorical, liturgical function. Or did it have an exclusively literary function as part of the Priestly narrative, with its subtle references to and connections with the Jerusalem temple symbolized by the tabernacle of Exod 25–40*, and will we have to limit ourselves to its *Sitz-in-der-Literatur*?[3] The general Ancient Near Eastern background of the

---

[1] H. Seebass, *Genesis I: Urgeschichte (1,1–11,26)*, Neukirchen-Vluyn 1996, 47 and 62: 'Für die Gesamtanlage hat man bisher nichts Vergleichbares gefunden'.

[2] *Enuma elish*, the myth about Marduk's rise to leadership of the gods during which he gained victory over Tiamat and reorganized the universe, was recited on the fourth day of Nisan during the New Year Festival in New Babylonian times (*ANET*, Princeton 1969[3], 60–72, 501–503; J. Bottéro and S.N. Kramer, *Lorsque les dieux faisaient l'homme: Mythologie Mésopotamienne*, Paris 1989, 602–697; *TUAT* III/4, Gütersloh 1994, 565–602; *COS* I, Leiden 1997, 390–402). It is unlikely, however, that it was composed with this particular aim in mind.

[3] J. Blenkinsopp, 'The Structure of P', *Catholic Biblical Quarterly* 38 (1976) 275–292; M. Weinfeld, 'Sabbath, Temple and the Enthronement of the Lord—The Problem of the Sitz im Leben of Genesis 1:1–2:3', in: FS H. Cazelles (AOAT 212), Neukirchen-Vluyn 1981, 501–512; J. Levenson, *Creation and the Persistence of Evil*, San Francisco

connection between creation and cult, of the role of the temple as the centre of the cosmos, would favour such a concentration on the macro context. Or are both options part of an antiquated paradigm of alternatives no longer valid in the present debate on Genesis texts? The role of chronology in the Priestly Codex[4] may support such a vision; after all, the chronological markers may be used for *both* concepts. The world to be is brought into the dimension of time by the creation of light and the renaming of light *and* darkness as *day* and *night* (1:5). Chronologically, this means that the creation narrative starts on the first day of the first week of the first month of the first year. After the reversal of creation in the Great Flood the earth is freed from the waters on the first day of the first month (8:13).[5] This is a new beginning. The theophany at Sinai results in the command by YHWH to set up the tabernacle on the first day of the first month (Exod 40:2) and so Moses did (Exod 40:17).[6] From creation until the establishment of the sanctuary in the promised land, chronology bridges and connects the mighty acts of the creator of the world and the deity who wants to live in a tent in the midst of Israel.

But even within these well-known frameworks, the creation account leaves us with several question marks. Maybe the time for great con-

1988, 77–99; B. Janowski, 'Tempel und Schöpfung: Schöpfungstheologische Aspekte der priesterschriftlichen Heiligtumskonzeption' (1990), in: B. Janowski, *Gottes Gegenwart in Israel: Beiträge zur Theologie des Alten Testaments*, Neukirchen-Vluyn 1993, 214–246; B. Janowski, 'Der Himmel auf Erden: Zur kosmologischen Bedeutung des Tempels in der Umwelt Israels', in: B. Janowski and B. Ego (eds), *Das biblische Weltbild und seine altorientalischen Kontexte* (FAT 32), Tübingen 2001, 229–260.

[4] Though there is fierce debate on the usefulness of source criticism and on the size and character of the Priestly Codex, I think understanding the different voices in the Primeval Cycle Gen 1–11 in a source model is still helpful. Within the Primeval Cycle the following texts belong to the P-*Grundschrift:* Gen 1:1–2:4a; 5:1–28, 30–32; 6:9–22; 7:6, 11, 13–16a, 17a, 18–21, 24; 8:1, 2a, 3b–5, 13a, 14–19; 9:1–17, 28f.; 10:1–4a, 5*–7, 20, 22f., 31f.; 11:10–27*, 31f. For full coverage of the models and theories playing a role in the present (continental) debate, see C. Frevel, *Mit Blick auf das Land die Schöpfung erinnern: zum Ende der Priestergrundschrift* (HBS 23), Freiburg & New York 2000.

[5] Gen 7:(6), 11a, 23; 8:3b, 4, 5, 13, 14 result in a scheme of 354 days (12×29.5 lunar month days) from 17.II. 600* until 16.II.601*. The eleven days of 17.II.601–27.II.601 should be added. Then the duration of the flood in the Priestly version is 365 days, a year based on the solar calendar! For this theory see E. Kutsch, 'Der Kalender des Jubiläenbuches und das Alte und Neue Testament', *Vetus Testamentum* 11 (1961) 39–47, esp. 43 and C. Westermann, *Genesis* (BKAT I–1), Neukirchen-Vluyn 1974, 582f.

[6] Finally, the dedication of the Jerusalem temple (1 Kgs 8:2) is connected with the festivities in the seventh month, Ethanim, which is the New Year's Festival in autumn.

cepts—along the lines of those proposed during several decades of the last century—has still not come. What we have to offer here are some minor observations on the question of whether the Priestly Codex offers more than a sophisticated tale of the beginning of the world and humankind. Did these Priestly writers want to say something about Elohim himself, not only *how* everything came into being in their eyes, but by *whom*?

Before going into details, I would like to make two general remarks:

(1) Gen 1 is the most sophisticated account of creation in the Hebrew Bible. Not the Yahwistic narrative of Gen 2:4bff., with its vivid and colourful style, nor the Ancient Near Eastern view of creation as the battle between the creator and the chaotic powers of waters and monsters (Isa 51:9f.; Ps 74:12ff.; Job 26:10ff. etc.), nor the vision of Lady Wisdom as the mediatrix of creation at the beginning of the world (Prov 8:22ff.) demonstrate such a polished, balanced style as Gen 1.

(2) This text should be studied as a unity. Neither the differences between a 'Tat-' and 'Wort-Bericht'[7] nor the irregularity between the eight works of creation and the scheme of the seven days,[8] but rather the peculiar style[9] of this creation account is the starting point for the exegesis. The elaborate study by Steck opens with the statement: 'Der Schöpfungsbericht der Priesterschrift … erweckt wie kaum ein zweiter alttestamentlicher Text den Eindruck eines bis ins einzelne durchgeplanten, mit stereotypen Konstruktionselementen errichteten Ganzen, das auf einen einheitlichen Gestaltungswillen schliessen lässt'.[10] And von Rad formulated a similar point of view in view of the theological doctrine: 'Nothing is here by chance; everything must be considered carefully, deliberately, and precisely … These sentences cannot be easily over interpreted theologically! Indeed, to us the danger appears greater

---

[7] W.H. Schmidt, *Die Schöpfungsgeschichte der Priesterschrift: Zur Überlieferungsgeschichte von Genesis 1,1–2,4a und 2,4b–3,24* (WMANT 17), Neukirchen-Vluyn 1967².

[8] Schmidt, *Die Schöpfungsgeschichte der Priesterschrift*, 54–56.

[9] S.E. McEvenue, *The Narrative Style of the Priestly Writer* (AnBib 50), Rome 1971; B.W. Anderson, 'A Stylistic Study of the Priestly Creation Story', in: G. Coats and B. Long (eds), *Canon and Authority*, Philadelphia 1977, 148–162; G. Wenham, *Genesis 1–15* (WBC), Waco 1987.

[10] O.H. Steck, *Der Schöpfungsbericht der Priesterschrift: Studien zur literarkritischen und überlieferungsgeschichtlichen Problematik von Genesis 1,1–2,4a* (FRLANT 115), Göttingen 1981², 11.

that the expositor will fall short of discovering the concentrated doctrinal content'.[11] These might be dangerous sentences, allowing everybody to find everything, but von Rad is right. Therefore the aim of this paper is to highlight one aspect of this concentrated priestly doctrine: the role of light in relation to darkness in the first work of creation.

Three steps are necessary for such an investigation:

1. What makes the creation of light and the naming of day and night so special in relation to the whole of the creation account?
2. What is the relationship between vv. 3–5 and v. 2 and v. 1?
3. What is meant by the common and still exceptional combination of light and darkness in vv. 3–5?

1. *What makes Gen 1:3–5 so special?*

(a) First of all, the simple observation that the P account separates the creation of light (vv. 3–5) from the creation of sun, moon and stars (vv. 14–18). Because the text is so familiar to many readers, we are accustomed to take it for granted. But this remarkable feature has only one parallel[12] in the Hebrew Bible, Ps 74:16 לך יום אף־לך לילה אתה הכינות מאור ושמש: 'Yours is the day, yours also the night, you established the light and the sun'. And even this parallel is not that certain for מאור, 'light', probably means 'luminary' here, and the same word is indeed used in Gen 1:14ff. for the creation of sun, moon and stars. It is the first indication that we are dealing with an intentional separation between the first work of creation and subsequent ones.

(b) Secondly, 'day' and 'night' come into being by separation.

V.	*Part of*	*Separation between*	*Instrument*	*Separation*
4	Realization	Light and darkness	Elohim	ויבדל
6	Command	Waters	רקיע	מבדיל
7	Realization	Waters and waters	Elohim	ויבדל
14	Command	Day and night	מארת	להבדיל
18	Realization	Light and darkness	מארת	ולהבדיל

[11] G. von Rad, *Genesis: A Commentary* (OTL), Philadelphia 1973, 48f. = *Das erste Buch Mose: Genesis* (ATD 2–4), Göttingen 1972[9].

[12] Schmidt, *Die Schöpfungsgeschichte der Priesterschrift*, 95–99.

The table above demonstrates that the concept of separation is common to Gen 1. The waters are separated by the 'dome', day and night, light and darkness are separated by the 'lights' of vv. 14–18. Looking back to Gen 1:4, it is striking that there is separation only where light is created and darkness already existed (v. 2). Nowhere, however, can a mixture of light and darkness be presupposed. All the signals of the text indicate that the light should be set apart.

(c) Thirdly, in v. 3 the light is the only object of Elohim's creation act, not the darkness.

(d) Finally, the enigmatic correspondence between יהי אור ויהי־אור (v. 3) is not found elsewhere. Though Gen 1:6 starts with a יהי רקיע, the realization of the dome is told with ויעשׂ אלהים, not with ויהי. The same is the case with the creation of the lights in v. 14. The divine command יהי מארת is followed by a different verb: ויעשׂ אלהים (v. 16). Jacob observed that there is a difference between all the verbs used for commands on the one hand and their execution on the other, not just for X- יהי· vv. 9–10, 11–12, 20–21, 24–25, 26–27.[13]

Examining the study by the late Odil Hannes Steck[14] in detail is very helpful. What most commentaries or monographs state in general terms has been meticulously worked out by him. I shall mention the most important observations. First, there is the question of where and how the formula of approval is used in the creation account.

V.	*Day*	*No.*	*Work*	*Formula of Approval*
4a	1	1	light > day [+night]	וירא אלהים את־האור כי־טוב
8	2	(2)	LXX: dome	LXX: καὶ ἐκάλεσεν ὁ θεὸς τὸ στερέωμα οὐρανόν καὶ εἶδεν ὁ θεὸς ὅτι καλόν
10b	3	3	dry land > earth [+sea]	וירא אלהים כי־טוב
12b	3	4	vegetation < earth	וירא אלהים כי־טוב
18b	4	5	lights > two great lights and the stars	וירא אלהים כי־טוב
21b	5	6	sea animals + birds	וירא אלהים כי־טוב
25b	6	7	earth animals	וירא אלהים כי־טוב
31	6	-	concluding formula of approval	וירא אלהים את־כל־אשר עשה והנה־טוב מאד

[13] B. Jacob, *Das erste Buch der Tora: Genesis*, Berlin 1934, 30f.

[14] Steck, *Der Schöpfungsbericht der Priesterschrift*, 158–177.

In addition to 1:3–5, six further creation works are approved by Elohim in MT, making a total of seven: vv. (4), 10, 12, 18, 21, 25, 31. LXX adds the formula in v. 8, breaks through the sevenfold use of MT and reaches a standardization because now each day's work has its own formula. This happens, however, at the cost of content, for heaven is not yet completed.[15] Therefore it is unlikely that the formula in v. 2 belongs to the original text. Generally speaking, Elohim looks at the result of his creation act and calls it 'good': וירא אלהים כי־טוב (see also Dehandschutter, this volume, §3). Twice the approval appears at the end of the creation act before the day formula ends the unit (vv. 12, 18). Twice the approval separates the first and the second work of creation on one day (vv. 10, 25). Once the approval appears before the blessing of the creatures (v. 21). In all five cases the formulation of the approval is the same. The final v. 31 with והנה־טוב מאד differs. But this final evaluation of creation reviews the totality of the six days. It is the *summa* of Elohim's work and deserves an extra: טוב מאד. V. 31 provides an understandable and expectable טוב מאד.

Viewed from this regular use of וירא אלהים כי־טוב as a monotonous phrase, the formulation of v. 4 וירא אלהים את־האור כי־טוב is exceptional indeed. Nowhere else in the creation narrative is the object itself included in the formula. Here Elohim sees that *the light* is good. The light *itself* receives the predicate טוב. With this slight change to the regular formula a signal is given. With this subtle reference the Priestly writer states that *Light* will have an extraordinary place in the creation story.[16]

Although this first observation is subtle but clear, more problematic is the place and the function of ויהי־כן in Gen 1:7, 9, 11, 14, 24, 30, usually translated by 'and it was so'. The formula is understood as a final phrase after the creative word of Elohim. Steck disputes this common view, referring to three texts outside Gen 1 (Judg 6:38; 2 Kgs 7:20; 15:12) and adding a fourth (Judg 6:40), and concludes that the formula links word and occurrence as a 'Feststellung folgerichtiger Entsprechung'.[17] This means ויהי־כן claims that the word of Elohim is carried out in such a way that the result is in conformity with that word and should be translated: 'and accordingly it happened …'

---

[15] See MT v. 18b.

[16] Steck, *Der Schöpfungsbericht der Priesterschrift*, 158ff.

[17] Steck, *Der Schöpfungsbericht der Priesterschrift*, 32–39.

Verse	*Day*	*No.*	*Work*	*Formula*	*Place of the Formula (F) in MT / LXX*
7b	2	2	expanse > firmament	ויהי־כן καὶ ἐγένετο οὕτως	**MT**: act of creation—F—naming **LXX**: 6b God's command—F—act of creation
9	3	3	waters > dry land > earth	ויהי־כן καὶ ἐγένετο οὕτως	**MT**: Elohim's command—F—naming **LXX**: God's command —F— **LXX plus**: καὶ συνήχθη τὸ ὕδωρ τὸ ὑποκάτω τοῦ οὐρανοῦ εἰς τὰς συναγωγὰς αὐτῶν καὶ ὤφθη ἡ ξηρά—indirect creation act
11	3	4	earth > vegetation	ויהי־כן καὶ ἐγένετο οὕτως	**MT**: Elohim's command to the earth—F—indirect creation act[18] of the *earth*//**LXX**
15	4	5	lights and the stars	ויהי־כן καὶ ἐγένετο οὕτως	**MT**: Elohim's command—F—Elohim's act of creation//**LXX**
20b	5	6	water animals and birds	καὶ ἐγένετο οὕτως	**LXX**: God's command—F (LXX plus)—God's act of creation
24	6	7	earth animals	ויהי־כן καὶ ἐγένετο οὕτως	**MT**: Elohim's command—F—Elohim's act of creation//**LXX**
30	6	8	gift of nourishment	ויהי־כן καὶ ἐγένετο οὕτως	**MT:** Divine speech about nourishment—F—approval of the whole of creation//**LXX**

The table above shows the place and the connections with the divine speech on the one hand and the creation acts on the other. To maintain his own scheme and to explain the functional place of the formula, Steck has to explain the variations of 2 Kgs 7:20 and Judg 6:40, to transfer ויהי־כן from v. 7a to v. 6,[19] to add the formula at the end of v. 20,[20] to presuppose that between the command and formula in v. 9 the act of creation itself is omitted at the beginning of v. 10,[21] and to transfer the formula from v. 7b to the beginning of v. 7a.[22] This results in a fitting scheme and Steck claims that ויהי־כן now appears in 1:6–25 in every creation work and, what is more, in the right place. In MT, however, only vv. 11, 15 and 24 fit his scheme. V. 11 is a special case

[18] K.A. Mathews, *Genesis 1–11:26* (New American Commentary 1A), Nashville 1996, 152 refers to 1:20, 24 for further examples of Elohim's indirect creative decree. Both verses are followed, however, by ויברא אלהים and ויעשׂ אלהים. This is different in v. 11f.

[19] BHK and BHS support the proposal (LXX).

[20] Again supported by BHK, BHS and LXX.

[21] V. 10 already starts with Elohim giving the dry land its name.

[22] Steck, *Der Schöpfungsbericht der Priesterschrift*, 40–44.

because the formula is not followed by a work of Elohim, but by an act of the earth as co-creator. With the different place of the formula in v. 6 LXX, the addition in v. 9 LXX, and the addition of the formula in v. 20 LXX Steck adds three more texts to his scheme. This means the whole construction leans heavily on the LXX of Genesis 1 as a better and original *Vorlage* of MT. The differences between MT and LXX in the first chapter of Genesis, however, demonstrate the systematic approach of the Greek and cannot be used automatically to 'improve' the Hebrew text.

A systematic analysis of the Greek version has been offered by Rösel.[23] He discusses the important studies by Cook[24] and Görg[25] as well as the proposals made by Harl[26] in her recent translation, all of which are relevant to our theme. Therefore I shall refer only to the necessary points in the discussion with Steck. The Greek translator offers a very detailed and careful translation of the Hebrew text. Sometimes he uses vocabulary which cannot be traced back to the MT. In other cases he clearly tries to reorder the text and to systematize it. LXX systematizes the use of the verbs used for creation: ποιέω covers both ברא (1:1, 21, 27 [3x]) and עשׂה (1:7, 16, 25, 31). LXX offers with its twofold καὶ ὤφθη ἡ ξηρά (v. 9) a conscious connection with v. 2 LXX ἡ δὲ γῆ ἦν ἀόρατος and introduces a logic for *invisible* and *visible* which is absent from MT (see also Dillon, this volume, §2, and Van Kooten, §1.2a). Here the translator realized that the earth becomes *visible* in v. 9, which means that it had to be invisible in v. 2. The special translation of v. 2 was followed by the plus καὶ ὤφθη ἡ ξηρά. Given the character of the translation, there is no proof that the translator used a different *Vorlage* of the Hebrew text. The conclusion must be that reconstructions following LXX are a dangerous undertaking in this part of Genesis.

There is some variation in the use and place of ויהי־כן when we do not want to follow Steck in his reconstructions. In v. 7, ויהי־כן appears after the act of creation and before the naming of the sky. V. 9 places ויהי־כן between the divine command and the naming of earth and sea.

[23] M. Rösel, *Übersetzung als Vollendung der Auslegung: Studien zur Genesis-Septuaginta* (BZAW 223), Berlin & New York 1994 (PhD thesis Hamburg 1993).

[24] J. Cook, 'Genesis 1 in the Septuagint as Example of the Problem: Text and Tradition', *Journal of Northwest Semitic Languages* 10 (1982) 25–36.

[25] M. Görg, 'Ptolemaische Theologie in der Septuaginta', *Kairos* 20 (1978) 208–217.

[26] M. Harl, *La Genèse* (La bible d'Alexandrie—LXX), Paris 1994[2].

Most often, ויהי־כן finds a place between the divine command and the realization or completion of the work. This is how it is in vv. 11, 15 and 24. V. 30 uses it between the gift speech by Elohim and the approval of the whole of creation.

If we leave the different positions aside for a moment, it is clear that the formula is missing in verses 3–5. The reason why is not totally clear, but something can be said. It is impossible to give ויהי־כן a place between command and completion because of the close connection between ויהי ... יהי. To put it in the same place as the first (v. 7) or second (v. 9) exceptions to the scheme would not work either because of the stress on the connection between ויהי ... יהי on the one hand and the unique formula of approval used in v. 4 on the other. The conclusion must be that it is indeed remarkable that ויהי־כן is missing, but that the other textual signals are so strong that there was no need for the formula.

To sum up, the first work of creation, light, is in a category of its own in structure, form and content. It is set apart from the creation of the luminaries. The creation of light in relation to the already existing darkness and their subsequent separation differs from the other works where separation plays a role. Only the light is created by word and realized by the same verb, without a different verb of completion. The formula of approval is the only one where the object of creation is named. The result of the last two remarks is that there is no other work of creation where the creator, Elohim, and the created object, light, are brought together this closely. The missing ויהי־כן formula confirms the careful way in which the specific elements of the first work of creation are intermingled.

2. *The relation of vv. 1 and 2 to the first work of creation vv. 3–5*

The problem of reading Genesis starts with the very first word בְּרֵאשִׁית.[27] The traditional position sees a construct in the vocalisation *b*ᵉ- whereas the absolute should read *bā*- using the definitive article. The understanding of the syntax of v. 1 (and 1–3) moves along two main lines.

---

[27] N.H. Ridderbos, 'Gen 1:1 en 2', *Oudtestamentische Studiën* 12 (1958) 214–260; P. Humbert, 'Encore le premier mot de la bible', *Zeitschrift für die alttestamentliche Wissenschaft* 76

V. 1 is a temporal clause 'In the beginning when Elohim created' followed by the main clause either in v. 2 'The earth was ...' or in v. 3 'Elohim said'.[28] The second possibility is to translate v. 1 as an independent main clause 'In the beginning Elohim created the heaven and the earth'. This is the most popular translation.[29] Returning to the presumed construct state of בְּרֵאשִׁית, several positions can be found: (a) it is a construct, but nevertheless it can be translated as an absolute main clause; (b) it is not a construct at all due to the fact that 'temporal phrases often lack the article'.[30]

What looks like a technical problem for linguists at first sight turns out to be a major theological problem in the history of reception. Three fields are touched on here: (a) the concept of a *creatio ex nihilo* (see also Tieleman, this volume, §1) absent from the text of Gen 1 but read into it from the times of 2 Macc 7:28 on: 'look at the sky and the earth; see all that is in them and realize that God made them out of nothing (ὅτι οὐκ ἐξ ὄντων ἐποίησεν αὐτὰ ὁ θεός)';[31] (b) the instrumental understanding of ב in בראשית relating it to 'wisdom'[32] (Prov 8:22) or 'torah' (Sir 24:23) in Jewish tradition, and (c) the instrumental understanding in some Christian traditions of ראשית as 'son' due to a wordplay on the first two letters of בראשית and Aramaic בר 'son'. In the first case, the verse at the very beginning was understood as 'By means of wisdom/torah Elohim has created heaven and earth', in the second 'By means of (the)

(1964) 121–131; P. Schäfer, 'Zur Interpretation von Genesis 1:1 in der rabbinischen Literatur', *Journal for the Study of Judaism* 2 (1971) 161–166; K. Deurloo and R. Zuurmond, '"In den beginne" en de "adem Gods" (Genesis 1:1,2)', *Amsterdamse Cahiers voor exegese en Bijbelse theologie* 7, Kampen 1986, 9–24; E. Jenni, 'Erwägungen zu Gen 1,1 "am Anfang"', *Zeitschrift für Althebraistik* 2 (1989) 121–127; U. Rüterswörden and G. Warmuth, 'Ist *bršyt* mit Artikel zu vokalisieren?', in: W. Zwickel (ed.), *Biblische Welten: Festschrift M. Metzger* (OBO 123), Freiburg & Göttingen 1993, 167–175; M. Bauks, *Die Welt am Anfang: Zum Verhältnis von Vorwelt und Weltentstehung in Gen 1 und in der altorientalischen Literatur* (WMANT 74), Neukirchen-Vluyn 1997, 93–98.

[28] E.g. NEB 1970 noting that an absolute clause is possible too; TNK 1985; NRSV 1989.

[29] Dutch: StV 1637, Lutheran Translation 1750,1994; NBG 1951, NBV 2004 noting that the translation 'In the beginning, when God created heaven and earth'—v. 2 'God said: ...' is possible too; German: Einheitsübersetzung 1980, Luther 1984; English: KJV 1611/1769; JPS 1917; RSV 1952; NJerBible.

[30] Wenham, *Genesis*, 12 mentions Gen 3:22, 6:3, 4; Isa 40:21, 41:4, 46:10 etc. Cf. the history of research: Westermann, *Genesis*, 131–135.

[31] The supporters of a *creatio ex nihilo* refer to Wisdom 11:17 ἡ παντοδύναμός σου χεὶρ καὶ κτίσασα τὸν κόσμον ἐξ ἀμόρφου ὕλης—'out of formless matter'.

[32] Cf. TgJ Gen 1:1.

son Elohim has created heaven and earth.'[33] With such a theologically loaded history of reception, it is difficult to find the way back to the position of the Priestly writer. Throughout the history of interpretation, theological rather than philological arguments have played the most important role. With regard to the latter, Rüterswörden and Warmuth made a critical survey of the text-critical arguments normally used in the field because most commentators have taken the apparatus of BHK and BHS for granted, presuming that a *ba-* transcription is the result of a *Vorlage* with the definite article and there is enough textual proof for the *ba-* reading among the Fathers.

The manuscripts reading βαρησηθ or βαρησειθ mentioned in the apparatus of the Göttingen Septuagint, however, only go back to the eleventh or twelfth century. Field does not add any extra material and the reference concerning BHK and BHS to the hexaplaric reading of Origen βαρησηθ cannot be verified. The value of the Samaritan reading *bàrášit* is limited. On the other hand, the second reference, βρησιθ, βρασιθ, is found several times in the Fathers, so the conclusion may be that the reading βα- is not a text-critical one but rather a transcriptional variant, and βρ is the normal one.[34] It can be demonstrated that βα- is used for a transcription of Hebrew ב with *shewa mobile*. This means that the old contrast between an absolute and a construct state of בראשית is without real proof. There is no construction *ba-* demonstrating an original reading with the definitive article. This, however, does not solve all the problems.

The next question is how to translate the original בְּרֵאשִׁית? Should it be a main clause or a dependent temporal phrase? For the first possibility the classical arguments are still valuable.[35] (1) The Septuagint translates in an absolute way, demonstrating that the reading of the text as a main clause was already present as early as the third century BC. (2) The Massoretes favoured the same reading by placing a distinctive *tifcha* under בראשית. (3) John 1:1 Ἐν ἀρχῇ ἦν ὁ λόγος refers to a reading of Gen 1:1 as a main clause. (4) A reconstruction of a clause running from

[33] For the details of the different readings of the Targumim and the positions of Rashi, Ibn Ezra on the one hand and Hilarius and Jerome on the other, see the articles by Deurloo & Zuurmond, '"In den beginne"' and especially Schäfer, 'Zur Interpretation von Genesis 1:1' and Rüterswörden & Warmuth, 'Ist *bršyt* mit Artikel zu vokalisieren?'

[34] Rüterswörden & Warmuth, 'Ist *bršyt* mit Artikel zu vokalisieren?', 172.

[35] Westermann, *Genesis*, 130–137.

v. 1–3 would not fit in with the style of the Priestly Writer. (5) Prov 8:22+Sir 15:14 prove that בְּרֵאשִׁית can be used without an article. (6) רֵאשִׁית can be understood in an absolute way.[36]

In the recent discussion about the syntactic construction of v. 1, it has become clear that the reading בְּ is the better one. This, however, does not mean that a construct state should be solved by reconstructing a long dependent clause composed of the elements of vv. 1–2 or 1–3. On the contrary, on the basis of the arguments above, an absolute translation is to be preferred. Bauks comes to the same conclusion, referring to v. 1 as an anacoluthon and translating: 'Im Anfang (war es), als Gott die Welt[37] schuf.'[38] Although uncertainties remain, there is no compelling reason to translate v. 1 in another way than 'In the beginning Elohim created heaven and earth', understanding it as the heading and summa of the whole Priestly account of creation. This makes vv. 2 and 3 separate verses with the act of creation starting in v. 3.

Benno Jacob's explanation is too beautiful not to be mentioned here. He also translates בְּרֵאשִׁית as an absolute, observing that the whole account of creation starts with *waw-consecutivum* clauses. This, however, must be made possible by an extraordinary beginning of everything. How extraordinary can be demonstrated by comparing Gen 1:1 with texts which also speak of the beginning:

> Um die Neuheit und die Kraft dieses בְּרֵאשִׁית zu würdigen, sehe man, wie spätere Schriftsteller darum ringen, den Begriff der unzeitlichen Vorweltlichkeit zu verdeutlichen, z.B. Prov 8,22vv durch nicht weniger als zehn verschiedene Ausdrücke … Dass בְּרֵאשִׁית in diesem absoluten Sinne nicht wieder vorkommt, ist nicht bloss damit zu erklären, dass es eben nie wieder einen vorzeitlichen Anfang gegeben hat … vielleicht wollte man es als ausschliessliches Eigentum der Tora und ihres Anfanges respektieren.[39]

Two extremes should be avoided for the function of v. 2. Neither the description of the earth as תהו ובהו, the תהום, or the ורוח אלהים מרחפת

[36] Isa 46:10; cf. Prov 8:23.

[37] Bauks, *Die Welt am Anfang*, 145 wants to stress the world as the place of created life and replaces the merism of heaven and earth by the expression 'world'. Because of the use of heaven and earth in the following verses, I prefer to maintain the merism.

[38] Bauks, *Die Welt am Anfang*, 145–146.

[39] B. Jacob, *Das erste Buch der Tora: Genesis*, Berlin 1934, 21f.

על־פני־המים are material with which or out of which creation occurred, nor is it the monstrous chaos which was conquered by the battle powers of the creator god. תהום is silent here. It has only the etymology in common with the battle monster from *Enuma elish*, nothing more.

V. 2 aims at contrast and functions as a counterpart to the coming creation. Here, Bauks' description avoiding extreme positions is helpful: 'Vielmehr ist Gen 1,2 auf dem Hintergrund einer verbreiteten Schöpfungstradtion zu verstehen, die anstatt das der Schöpfung Vorausgehende detailliert zu beschreiben, es in Verkehrung zur gegenwärtigen Welt in einem Kontrastbild darstellt.'[40] That this counterpart is used with the aim of limitation, however, we will see later on with the antipodes 'darkness-light'.

One problem remains. Is v. 2 pure contrast or already a stepping stone to v. 3? This depends on the interpretation of the last part of v. 2, especially רוח אלהים. Translations governed by denominational interests have interpreted the רוח אלהים as the Spirit of God. This is still the most common translation.[41] When biblical interpretation became aware of the cosmogonies of the world around Israel, the translation of wind or storm[42] grew popular, taking אלהים as a superlative: a *mighty* storm/wind.[43] Although philologically defendable, it does not, however, fit into the language and content of the creation account. In this dense text Elohim plays a dominant role in speech and actions. The use of אלהים as an adjective is inappropriate here.[44] The translation 'wind of God' or 'breath of God' receives more support.[45] The difficulty with

---

[40] Bauks, *Die Welt am Anfang*, 144.

[41] LXX, cf. Rösel, *Übersetzung als Vollendung der Auslegung*, 34f. Harl, *La Genèse*, 85 translates: 'le souffle de Dieu', cf. 87 for her explanation; Vulgate; Modern translations: SV, NBG, NBV, LuthV.; German: Zürcher; Luther 1984; Einheitsübersetzung; English: RSV, NKJames.

[42] W.H. Schmidt, *Die Schöpfungsgeschichte der Priesterschrift* (WMANT 17), Neukirchen-Vluyn, 1973[3], 83.

[43] Westermann, *Genesis*, 107 ('Gottessturm'), 149.

[44] In the three groups where רוח אלהים is used in the Hebrew Bible: (1) Gen 41:38; Exod 31:3, 35:31 (gifted skill); (2) Num 24:2; 1Sam 10:10, 11:6, 19:20, 23; 2 Chr 15:1 ([ecstatic] prophetic ability); (3) 1Sam 16:15, 23; 18:10 (evil spirit), אלהים means always 'God'.

[45] Already defended by Ridderbos, 'Gen 1:1 en 2', 245 referring to Ps 33:6: 'Erst geht der Atem Gottes über die Wasser: er zügelt die Wasser und die Finsternis. Dann wird aus dem Atem Gottes ein Sprechen und auf das Wort Gottes muss die Finsternis fliehen und die Wasser erhalten ihren Platz angeweisen.' Along the same line and arguments but in a different setting: Steck, *Der Schöpfungsbericht der Priesterschrift*, 236.

wind, however, is the verb רחף, which does not fit in the context of wind or storm, so most newer commentaries favour the solution 'breath'.[46]

This last solution may be questioned by having a look at the texts of the Hebrew Bible where the רוח אלהים means 'Spirit of God'. The Yahwistic narrative of the creation of man and woman (Gen 2:4bff.) sees the beginning of life as the divine action of blowing the breath of life into the nostrils of man. From that point on man becomes a living creature (v. 7). The words used for 'breath of life' are נשמת חיים. The same type of (re)creation in the later text Ezek 37 is expressed by רוח.[47] There is certainly a wordplay between 'wind' and 'spirit,' but it is clear that the life-giving spirit of YHWH is called רוח in the text of Ezekiel (v. 14). Up until Qoheleth, the substantive will be used in this way. The creation account of Gen 1 stands in the same priestly tradition as Ezekiel, taking over the רוח אלהים from Ezekiel but giving it a different place. From v. 3 on, everything is ruled by the divine word and even the רוח אלהים does not have a place there. Therefore it finds its position at the end of v. 2, bridging the world of contrast in the first parts of v. 2 with the divine action and word of v. 3. The powers of death, water and darkness have already been breathed upon by the Spirit of Elohim. Along these lines, רוח אלהים already had a long history in the Hebrew bible before it was allowed to act in the Priestly account of the very beginning.

3. *Light and darkness in Gen 1:3–5 as attributes of Elohim*

As long as Gen 1 was studied from an exclusively cosmogonic point of view, the relation between vv. 3–5 and 14–18 was one of the main items (see also Tigchelaar, this volume, *passim*).[48] However, it is not

---

[46] Wenham, *Genesis*, 2; H. Jagersma, *Verklaring van de hebreeuwse bijbel: Genesis 1:1–25:11*, Nijkerk 1995, 23; Seebass, *Genesis I: Urgeschichte*, 58 ('Hauch Gottes'); Bauks, *Die Welt am Anfang*, 146 ('Windhauch Gottes'). For the argumentation pattern see E. Noort, 'Tod und Zukunft. Das Wagnis des Ezechiel: Ez 37,1–14 und die eschatologische Hoffnung', in: E. Noort and M. Popovic (eds), *Hoffnung für die Zukunft: Modelle eschatologischen und apokalyptischen Denkens* (Vorträge der zweiten Konferenz der Südostmitteleuropaïschen und Niederländischen Theologischen Fakultäten in Cluj, Rumänien 2000), Groningen 2001, 7–16.

[47] Especially vv. 5, 8, 9, 14.

[48] Because the light before sunrise and the morning was regarded as the mytholog-

the question of how it was possible to create light without the sun and the other celestial bodies that is the important problem. More interesting is the question of whether and how vv. 3–5 speak about Elohim. For the relation with vv. 14–18 it is sufficient to note that 14–18 presuppose the rhythm of time, of day and night, made possible by the first verses of the creation account. Speaking about God in relation to light and darkness is never a neutral theme. Nor was it in the world predating and surrounding the time of the Priestly writer. The majestic presence of God is described in categories of light. In the Priestly Document this is reflected by the shining face of Moses when he descended from Sinai: 'Moses did not know that the skin of his face shone because he had been talking with God' (Exod 34:29). The idea is that Moses went up into the cloud with the כבוד־יהוה (Exod 24:15–18). In this direct confrontation with the 'glory' of God he automatically becomes a light-bearer too.[49] Before the Priestly Document, it was again Ezekiel who connected the כבוד־יהוה with the light of God's theophany (Ezek 1:26–28). In the hymns of the Hebrew Bible it is said: 'You are clothed with honour and majesty, wrapped in light as with a garment.'[50] In common with the whole of the Ancient Near Eastern world, light and darkness are used as metaphors for happiness and life, for misfortune and death.

Where and how does the pair אור and חשך play a role in the Hebrew Bible? In addition to Gen 1:4–5, 18 there is the book of Job. In Job 26:10, God's cosmic and creative power is demonstrated: 'He has described a circle[51] on the face of the waters, at the boundary between light and darkness.' The reminiscence of creation as a battle between God and the chaos monster Rahab[52] refers to the boundary set at cre-

---

ical time of help from the deity, light could be unlinked from the light of the sun: Ps 65:7–9. See also Ps 74:13–17.

49 Contra S. Aalen, 'אור', in: *Theologisches Wörterbuch zum Alten Testament* I (1971), 160–182, esp. section IV.1, 178.

50 Ps 104:2. Sometimes God himself can be designated as 'light': Ps 27:1, 2 Sam 22:29, Isa 10:17, 60:1, Mic 7:8. For Light as an attribute of God, see S. Aalen, *Die Begriffe 'Licht' und 'Finsternis' im Alten Testament, im Spätjudentum und im Rabbinismus* (Skrifter der Norske Videnskaps-Akademi Oslo), Oslo 1951; T. Podella, *Das Lichtkleid JHWHs: Untersuchungen zur Gestalthaftigkeit Gottes im Alten Testament und seiner altorientalischen Umwelt* (FAT 15), Tübingen 1996.

51 The horizon between the waters surrounding the earth and the firmament is meant.

52 Job 26:12, cf. 9:13.

ation between light and darkness. There is no preference for one or the other. The focus is the 'thunder of God's power' (Job 26:14), not the quality of light or darkness. The same applies to Job 38:19. Here Job is asked whether he knows the way to the dwelling of light and the place of darkness so that he may use it in the way God does to govern creation.[53]

After Job 12:22, God is empowered to bring the darkness to light: 'He uncovers the deeps of the darkness and brings deep darkness to light'. Even if Fohrer[54] is right that v. 22 is a gloss, the glossator uses this imagery of light and darkness to give God power over the evildoers on earth. Light is the place for the upright, they will rise in the darkness as a light (Ps 112:4). Light is the image of the salvation by YHWH: 'I will lead the blind by a road they do not know, by paths they have not known. I will guide them. I will turn the darkness before them into light, the rough places into level ground' (Isa 42:16).

Already from an early time, the light-darkness oppositions are not only used in the context of creation. Light and darkness become synonyms for welfare, salvation, good and misery, misfortune, evil. It is already used in this way—though located in the cosmic setting of the day of YHWH—in Amos 5:18, 20:

> Alas for you who desire the day of YHWH!
> Why do you want the day of YHWH?
> It is darkness, not light! (הוא־חשך ולא־אור) ... (18)
> Is not the day of YHWH darkness, not light
> And gloom with no brightness in it?[55] (20)

It is referred to in the harsh statements of Isa 5:20 primarily in an ethical way:

> Ah, you who call evil (רע) good (טוב)
> and good (טוב) evil (רע).
> Who put darkness (חשך) for light (אור)
> And light (אור) for darkness (חשך).

And the same imagery is still used at the other end of the Isaiah tradition:

[53] Cf. Job 38:22–24 where hail and snow are reserved for the days of battle and war.
[54] G. Fohrer, *Hiob* (KAT), Gütersloh 1963, 237.
[55] Cf. Ps 139:11f. and Lam 3:2.

> Therefore justice (משפט) is far from us,
> and righteousness (צדקה) does not reach us;
> we wait for light (אור) and lo! there is darkness (חשך)
> and for brightness, but we walk in gloom (Isa 59:9).

The Hebrew Bible is very reluctant to commit itself on the cause of evil. The snake suddenly appears in the Garden of Eden without any reason or explanation, it is simply there. The Priestly Document gives some reasons for the Flood,[56] but they are the result rather than the cause of man's evildoing. The authors of the Saul narrative have great difficulties explaining why Saul ends as he does. In the end they have to introduce an 'evil spirit' from YHWH to explain Saul's misfortune. These examples can easily be expanded. 'Unde malum' is not an item in the Hebrew bible. There is only one text which explicitly states that YHWH is the cause of evil: Isa 45:7.

> I form light (יוצר אור) and create darkness (ובורא חשך),
> I make well (עשׂה שלום) and create woe (ובורא רע);
> I, YHWH do all these things.

Here, the whole terminology of creation is concentrated. The important verbs יצר, ברא, עשׂה appear. What was noted above about the possibilities of light and darkness as part of creation, mixed up with ethical aspects and categories, is brought to an end in the enigmatic saying of Deutero-Isaiah. Light is שׁלום and darkness is רע. These are the texts before and around the Priestly account of creation. How do these texts see the relationship between Elohim on the one hand and light and darkness on the other?

We have seen that no work of creation in which Elohim and the object of creation are that close is realized by the same verb without a verb of completion. Expressis verbis *the light* is called 'good', a structure without parallel in the creation account. The lack of the ויהי־כן formula confirms the compact structure of the first work of creation. The darkness is already present, only 'light' is created. The darkness resulting from the contrast summed up in v. 2 is not created but limited. In the separation from the light and in the name-giving, Elohim demonstrates his power over darkness. Darkness is no longer boundless but is given its place in the rhythm of time. It is only 'night'. And night is always followed by the light of morning and day, the specific times of God's presence and help.

---

56 Gen 6:11 ('earth was corrupt and full of violence').

Anyone who reads Gen 1 with only the question of how the world came into being misses the careful and balanced way in which the Priestly writer wanted to say who the creator Elohim is. In the scheme of the Priestly Document, the revelation of the divine name YHWH belongs to the time of Moses (Exod 6:2ff.). What is said here, however, is that Elohim not only created the world and humankind, but also that he *wants* light.

Although darkness is not denied, Elohim limits it in his world. The Priestly author does not go as far as Deutero-Isaiah. Darkness/evil does not come from the hands of Elohim, he sets boundaries. But all the subtle differences observed in the first work of creation, hinting at his special attention to the light, demonstrate that the author wants to endow Elohim with the qualities of the later revealed name YHWH.

BACK TO CHAOS: THE RELATIONSHIP BETWEEN JEREMIAH 4:23–26 AND GENESIS 1

Jacques T.A.G.M. van Ruiten

Introduction

In this paper I will concentrate on Jeremiah 4:23–26, which is an oracle of judgement.[1] In it, a vision of the reversion of the creation into chaos is being described. It is followed by the reaction of YHWH (Jer 4:27–28). The vision has a place among those oracles concerned

[1] Cf. the following commentaries: J. Bright, *Jeremiah* (AB 19), New York 1965, 28–34; R.P. Carroll, *Jeremiah: A Commentary*, London 1986, 168–171; P.C. Craigie, P.H. Kelley, and J.F. Drinkard, *Jeremiah* (WBC 26), Dallas, Texas 1991, 80–82; B. Duhm, *Das Buch Jeremia* (KHAT XI), Tübingen & Leipzig 1901, 53–54; F. Giesebrecht, *Das Buch Jeremia* (HKAT III), Göttingen 1894 (1907²); W.L. Holladay, *Jeremiah, I II: A Commentary on the Book of the Prophet Jeremiah* (Hermeneia), Philadelphia 1986–1989, vol. I, 163–168; J.R. Lundbom, *Jeremiah 1–20* (AB 21A), New York 1999; W. McKane, *A Critical and Exegetical Commentary on Jeremiah, I–II* (ICC), Edinburgh 1986–1996, vol. I, 106–111; P.D. Miller, 'Jeremiah', in: L.E. Kelch (ed.), *The New Interpreter's Bible, VI* (Nashville 2001), 553–926, esp. 614–615; W. Rudolph, *Jeremia* (HAT 12), Tübingen 1958, 29–31; A. van Selms, *Jeremia, I* (POT), Nijkerk 1980, 92–93; J.A. Thompson, *The Book of Jeremiah*, Grand Rapids, Michigan 1980, 228–231; A. Weiser, *Das Buch des Propheten Jeremia, 1–25,13* (ATD 20), Göttingen 1952. Specific studies on Jer 4:23–26 (28) can be found in R. Althann, 'The Oracles of Jeremiah in North West Semitic Research', *Scriptura* 16 (1985) 17–28; A. Borges de Sousa, 'Jer 4,23–26 als P-orientierter Abschnitt', *Zeitschrift für die alttestamentliche Wissenschaft* 105 (1993) 419–428; V. Eppstein, 'The Day of Yahweh in Jeremiah 4,23–28', *JBL* 87 (1968) 93–97; M. Fishbane, 'Jeremiah IV 23–26 and Job III 3–13: A Recovered Use of the Creation Pattern', *Vetus Testamentum* 21 (1971) 151–167; K.M. Hayes, 'Jeremiah IV 23: *TOHU* without *BOHU*', *Vetus Testamentum* 47 (1997) 247–249; W.L. Holladay, 'The Recovery of Poetic Passage of Jeremiah', *JBL* 85 (1966) 404–406; W.L. Holladay, 'Style, Irony and Authenticity in Jeremiah', *JBL* 81 (1962) 44–54; R. Liwak, *Der Prophet und die Geschichte: Eine literar-historische Untersuchung zum Jeremiabuch* (BWANT 121), Stuttgart 1987, 240–242; D.C. Olson, 'Jeremiah 4.5–31 and Apocalyptic Myth', *Journal for the Study of the Old Testament* 73 (1997) 81–107; R.M. Paterson, 'Repentance or Judgment: The Construction and Purpose of Jeremiah 2–6', *Expository Times* 96 (1985) 199–203; J.A. Soggin, 'The "Negation" in Jeremiah 4:27 and 5:10a, cf. 5:18b', *Biblica et Orientalia* 29 (1975) 179–183 (cf. *Biblica* 46 [1965] 56–59); J.W. Vancil, 'From Creation to Chaos: An Exegesis of Jeremiah 4:23–26', in: F.F. Kearley, E.P. Myers, and T.D. Hadley (eds), *Biblical Interpretation: Principles and Practices—Studies in Honor of Jack Pearl Lewis*, Grand Rapids, Michigan 1986, 181–192; H. Weippert, *Schöpfer des Himmels und der Erde: Ein Beitrag zur Theologie des Jeremiabuches* (SBS 102), Stuttgart 1981, 49–54.

with the enemy from the north (Jer 4:5–10:25). The disclosed chaos evokes unmistakably the creation story of the book of Genesis. I will first deal with the structure of Jer 4:23–26 and its place in the literary context, especially its relation with Jer 4:27–28. I will then go into the intertextual relationship of this passage with Genesis 1.

1. *The structure of Jer 4:23–28*

The text of Jer 4:23–28 reads, in translation, as follows:

Jeremiah 4:23–28 (RSV, with slight modifications)

23 a	I looked on the earth, and lo, it was waste and void;[2]
b	and to the heavens, and they had no light.
24 a	I looked on the mountains, and lo, they were quaking,
b	and all the hills moved to and fro.
25 a	I looked, and lo, there was no man,
b	and all the birds of the air had fled.
26 a	I looked, and lo, the fruitful land was a desert,
b	and all its cities were laid in ruins,[3]
c	before YHWH,
d	before his fierce anger.

[2] The use of the expression תהו ובהו ('it was waste and void') has resulted in some debate. Most textual witnesses read two words with the Massoretic Text, the Peshitta (*twh wbwh*), the Vulgate (*vacuat erat et nihili*), and Aquila, Symmachus and Theodotion (κενὴ καὶ οὐθὲν). The Septuagint, however, reads only one word (οὐθέν). This could be an indication that the Hebrew *Vorlage* of the Septuagint only has one word. The second word was introduced into the text only later under the influence of Gen 1:2. Cf. Hayes, 'Jeremiah IV 23', 247–249. A metrical argument has also been put forward to prove that the original Hebrew text of Jer 4:23 has either בהו or תהו. Cf. Rudolph, *Jeremia*, 29–31. According to some οὐθέν could be the rendering of בהו. In their view, תהו ו was added later. Cf. P. Volz, *Der Prophet Jeremia*, Leipzig 1928[2], 51; Rudolph, *Jeremia*, 32. The phrase κενὴ καὶ οὐθὲν shows that Aquila, Symmachus and Theodotion equated οὐθέν with בהו. They added κενὴ καὶ as an equivalent of תהו ו. According to others, however, it is also possible that οὐθέν is the rendering of תהו. Cf. Duhm, *Jeremia*, 53; Weippert, *Schöpfer*, 50 (note 91); Hayes, 'Jeremiah', 247–249. In their view ובהו is a later addition. In my opinion, a third option is also possible. The expression οὐθέν in the Septuagint could be a compact rendering of both Hebrew words. Therefore, I would prefer to consider the Massoretic text as representing the original Hebrew text.

[3] Quite a number of the Hebrew manuscripts do not read נתצו ('they were laid in ruins'), but נצתו ('they were burned'). This seems to be the Hebrew *Vorlage* of the LXX (ἐμπεπυρισμέναι). The Peshitta (*'t'rq*) and the Vulgate (*destructae sunt*), however, reflect the reading of the Massoretic text (נתצו).

27 a For thus says YHWH:
b The whole land shall be a desolation;
c and I will make its destruction complete.[4]
28 a For this the earth shall mourn,
b and the heavens above be black;
c for I have spoken, I have purposed;
d I have not relented nor will I turn back.

The original unity of the passage Jer 4:23–38 is disputed. It is clear that a new passage starts in v. 23. As far as the content is concerned, the subject changes. A new passage also starts in v. 27 because a messenger formula is used: כי כה אמר יהוה ('For thus says YHWH'). The particle כי ('for') can be interpreted causally, and could therefore connect vv. 27–28 with the preceding verses. In that case the passage could be understood as a unity, in that the last two verses give the underpinning of God to what the prophet sees in the first three.[5] However, other arguments also point to the conclusion that vv. 23–26 and vv. 27–28 should be considered as separate unities.

As far as the formal structure of vv. 23–26 and that of vv. 27–28 are concerned, it is possible to differentiate between the two passages. The first part (Jer 4:23–26) consists of five lines, all of them bicola. Although it is possible that YHWH is the speaker in these verses, it is more probable that Jeremiah is the one who is speaking here. The use of YHWH in the third person in v. 26 makes this clear. After YHWH is the speaker in v. 22, he is again the speaker in vv. 27–28. In the verses in between, we have to make do with a report by a visionary. The first four lines start with 'I looked' (ראיתי) and continue in the first colon with 'and lo' (והנה). In the second colon, וכל ('and all') appears three times, only in v. 23 is it ואל. The tight structure of the passage confirms its unity, which is shown in the following scheme:[6]

[4] The translation reflects וכלה לה אעשה. The negation לא has probably superseded the personal pronoun לה. Cf. Rudolph, *Jeremia*, 36; Weippert, *Schöpfer*, 50 (note 92). There is no textual evidence for the deletion of לא. However, the use of לא is in complete contradiction with the rest of Jeremiah 4. It may have been inserted into the text to soften the vision of complete destruction. Cf. McKane, *Jeremiah, I*, 109. Soggin suggests that the א of לא originated by way of dittography. The original וכלה לאעשה ('I will certainly make the destruction complete') became in the way of textual transmission וכלה לא אעשה. See J. Soggin, 'La negazione in Geremia 4,27 e 5,10a, cfr. 5,18b', *Biblica* 46 (1965) 56–59.

[5] Weiser, *Jeremia*, 46. Cf. W. Holladay, who considers vv. 27–28 as an affirmation of the vision of vv. 23–26, although he does not consider the connection as original; see Holladay, *Jeremia, I*, 151.

[6] Cf. W.L. Holladay, 'The Recovery of Poetic Passage of Jeremiah', *JBL* 85 (1966)

23	ראיתי ... והנה ... אל ...	I looked ... and lo ... and to
24	ראיתי ... והנה ... וכל ...	I looked ... and lo ... and all
25	ראיתי והנה ... וכל ...	I looked and lo ... and all
26a	ראיתי הנה ... וכל ...	I looked and lo ... and all
26b	מפני ... מפני	before ... before

The poetic lines of 4:23–26 are of a gradually diminishing length. According to some exegetes, this reflects the decline of the creation.[7] The hearers are left with silence at the end. The repetition at the beginning can be considered as a kind of anaphora. 4:26cd also reflects anaphora: '*before* YHWH, *before* his fierce anger'.

The verb 'to look' has an object in v. 23 and v. 24, but none in v. 25 or v. 26. Moreover, vv. 23–24 deal with non-living nature (earth, heavens, mountains, hills), whereas vv. 25–26 deal with living matters (4:25–26: man, birds, fruitful land, cities). For these reasons, it is possible to divide the poem of Jer 4:23–26 into two strophes: I. 4:23–24 and II. 4:25–26.

The second part (Jer 4:27b–28d) consists of three bicola, and YHWH is the speaker. Despite the different outlook of both parts, I would like to highlight some elements in 4:27–28 that occur in vv. 23–26: 'earth' and 'heaven' (4:23; 4:27–28a), and the last verb in 4:26 ('lay in ruins'), that as far as the content is concerned occurs in the first word of the speech of YHWH, 'a desolation' (4:27b). Finally, I would like to point out the use of 'all' in 4:24b, 25b, 26b. The same Hebrew word כל is used as in 'the *whole* land/earth' (4:27b). However, the difference in style and content with regard to the first part is evident.

Finally, the references to Genesis 1, which I will deal with in detail later, occur only in vv. 23–26 and not in vv. 27–28. Therefore, it is possible to differentiate sharply between vv. 23–26 and vv. 27–28.[8]

2. *Jer 4:23–26 as a later addition*

Several exegetes question the originality of Jer 4:23–26 in the book of Jeremiah. Its cosmic atmosphere, in the midst of texts that refer to the historical reality of Judah and Jerusalem, is unique for Jeremiah.

404–406.

[7] Lundbom, *Jeremiah 1–20*, 358.

[8] According to Duhm, *Jeremia*, 54, the later verses are no more than water after the precious wine of the earlier verses.

According to these exegetes, both the formal features and the content point to an apocalyptically inspired redactor who added this fragment to the book.[9]

Borges de Sousa put forward the specific use of the literary genre to prove that Jer 4:23–26 is not originally from Jeremiah.[10] According to him, Jer 4:23–26 has a clear structure and belongs to the genre of a vision report. Every presentation of a vision is introduced by the verb **ראה**, without mentioning the receiver of it. In the first two presentations of the vision (vv. 23–24), the introducing verb is followed by the object of the vision. The actual presentation is introduced by **הנה**. With regard to this, a nominal clause is followed by a verbal one. Finally, there are two nominal clauses introduced by **מפנה**. The genre of a vision report occurs only in a few other places in the book of Jeremiah, i.e. in Jeremiah 1 (vv. 11–12 and vv. 13–14) and 24. According to Borges de Sousa, the realization of this literary form in these passages, however, is very different from that in Jeremiah 4. In Jeremiah 4, there is no narrative report, no dialogue between YHWH and the prophet, and no auditive event that interprets the vision, as in the other passages. Jer 4:23–26 only presents a vision, in poetic style. Because of this, Borges de Sousa considers it probable that Jer 4:23–26 is not a genuine text of Jeremiah.

It is possible to have some doubts about this argumentation. Although the observation that Jer 4:23–26 differs from Jer 1:11–12, 13–14; 24 is true, it is questionable methodology to compare one text with only two others, from a similar somewhat vague literary genre, and conclude that because this text deviates from the two others, it is not a genuine part of Jeremiah.

In my opinion, it is in fact difficult to distinguish between earlier and later texts in the book of Jeremiah. The book not only has a peculiar text-critical history, it also has a very complex redactional history. It is not concerned with logic and a coherent design. Often the book seems to be a mixed bag with no order in the different types of texts. There are collections of oracles, confessions, dialogues and liturgical compositions, but also historical, biographical and even autobiographical stories.[11] The messy plan of the book and the tensions

---

[9] Cf. Giesebrecht, *Jeremia*, 28; Volz, *Jeremia*, 50; Carroll, *Jeremiah*, 168–171.

[10] A. Borges de Sousa, 'Jer 4,23–26 als P-orientierter Abschnitt', *Zeitschrift für die alttestamentliche Wissenschaft* 105 (1993) 419–428, esp. 424–427.

[11] E.g. Lundbom, *Jeremiah 1–20*, 85.

within it indicate a gradual growth. Nevertheless, it is difficult to isolate passages and to decide whether they are earlier or later.[12] This diversity of material is also present in Jeremiah 4. However, I find it difficult to separate Jer 4:23–26 from the rest of chapter 4. The material seems to be quite well structured, for which I refer to the following scheme.[13]

Envelope Structure of Jeremiah 4

A Sion (4:6a)
B your land was a waste (4:7)
C your cities will be ruins / without inhabitants (4:7)
D sackcloth / lament and wail (4:8)
E has not turned back … the fierce anger of YHWH (4:8: כי לא שב … הרון אף יהוה ממנו)
F הנה + eagles (4:13: 'his horses are swifter than eagles')
G I hear the sound of the trumpet (4:19)
G' How long must I … hear the sound of the trumpet? (4:21)
F' הנה + 'all the birds of the air' (4:25)
E' the fierce anger of YHWH (4:26: מפני יהוה / מפני הרון אפו) … I will not turn back (4:28: לא אשוב)
B' the whole land shall be a desolation (4:27)
D' the earth shall mourn / the heavens are black (4:28)
C' all cities are forsaken / no man dwells in them (4:29)
A' Sion (4:31)

Jeremiah 4 is not only well structured, there are also many contacts between Jeremiah 4:23–26 and the literary context, not only within this chapter but also with other parts of the book. As far as the connections of Jer 4:23–26 within Jeremiah 4 are concerned, it is possible to point out contacts. Jer 4:25a ('there was no man') resembles 4:7 ('[cities] without inhabitants') and 4:29 ('no man dwells in them'); 4:25b ('all the birds of the air') resembles 4:13 ('eagles'); 4:26a ('fruitful land was a desert') resembles 4:7 ('your land a waste') and 4:27 ('the whole land shall be a desolation'); 4:26b ('its cities were laid in

---

[12] See, e.g., W. Thiel, *Die deuteronomistische Redaktion von Jeremia 1–25* (WMANT 41), Neukirchen 1973; Holladay, *Jeremiah, II*, 10–24; Lundbom, *Jeremiah 1–20*, 92–101.

[13] This scheme is a slightly altered adaption of D.C. Olson, 'Jeremiah 4.5–31 and Apocalyptic Myth', *Journal for the Study of the Old Testament* 73 (1997) 81–107, esp. 82–83. According to Olson, Jer 4:23–28 is an 'apocalyptic' poem that presupposes a body of Enochic mythological beliefs, that occurs also in the 'Book of Watchers' (*1 Enoch* 1–36).

ruins') with 4:7 ('your cities will be ruins') and 4:29 ('all cities are forsaken'); 4:26cd: the fierce anger of YHWH with 4:8 ('the fierce anger of YHWH').

As far as the connections between Jer 4:23–26 and other parts of Jeremiah are concerned, I need only refer to Jer 4:25b ('all the birds of the air had fled') and Jer 9:9 ('the birds of the air … have fled …'); to 4:26b ('all its cities were laid in ruins' [נתץ]) and Jer 1:10 ('to lay in ruins' [נתץ]); the figure of speech in which the first word of the line is repeated also occurs in Jer 50:35–37; 51:20–23.

I think these arguments are enough to show that the passage is very well integrated in the book of Jeremiah. Moreover, in my opinion the passage is not an apocalyptic passage. The author makes no distinction between cosmic and historical events. Here, too, the focus remains on the destruction of Judah and Jerusalem. It is a type of prophetic speech that also occurs elsewhere in the prophetic corpus, e.g., Isa 65:17–25, which also refers to the first chapters of Genesis.[14] In a certain way, this text can be considered as a counterpart of Jeremiah 4. It speaks about the creation of new heavens and a new earth, but this is focused very clearly on the creation of a new Jerusalem.

3. *The intertextual relationship between Jer 4:23–26 and Genesis 1*

In the course of the history of exegesis, several exegetes have highlighted in one way or another the relationship between Jer 4:23–26 and Genesis 1.[15] It serves as a counterpart to the first chapter of Genesis. It is possible to point out several similarities between both texts.

(a) The most obvious similarity with Genesis 1 concerns the words 'waste (= without form) and void' (Jer 4:23: תהו ובהו). This colloca-

[14] See O.H. Steck, 'Der neue Himmel und die neue Erde: Beobachtungen zur Rezeption von Gen 1–3 in Jes 65,16b–25', in: J. van Ruiten and M. Vervenne (eds), *Studies in the Book of Isaiah* (BETL 132), Leuven 1997, 350–365. Cf. also J.T.A.G.M. van Ruiten, 'Eve's Pain in Childbearing? Interpretations of Gen 3:16a in Biblical and Early Jewish Texts', in: G.P. Luttikhuizen (ed.), *Eve's Children* (TBN 5), Leiden 2003, 3–27, esp. 9–11.

[15] E.g., Fishbane, 'Jeremiah IV 23–26', 151–154; Weippert, *Schöpfer*, 51–54; Holladay, *Jeremiah, I*, 164; Craigie, *Jeremiah*, 81; Lundbom, *Jeremiah 1–20*, 357. Carroll, *Jeremiah*, 168, and McKane, *Jeremiah, I*, 106–108 dispute that Jer 4:23–26 is dependent on Genesis 1.

tion only occurs here and in Gen 1:2. In addition, both words also occur in Isaiah 34:11, but in a different syntactical construction. The word בהו: 'void' occurs only in these places, whereas the word תהו: 'waste' occurs in other places as well. Jeremiah is viewing a complete chaos, a destruction of creation. And because of the use of these two words, it is likely that the text is alluding to the creation story in Genesis 1.

(b) There is, moreover, also the similarity between the phrase 'and they had no light' (Jer 4:23b) and the 'light' that God created on the first day (Gen 1:3: 'And there was light'). In Gen 1:3 the creation of the light is not explicitly connected with the heavens; however, this takes place on the fourth day, when the lights, the sun, moon and stars are created to give light on earth (Gen 1:14–19; see also Tigchelaar, this volume, *passim*).

(c) Other elements strengthen the intertextual relationship of Genesis 1:1–2:4a and Jer 4:23–28. I only need to point to the fourfold repetition of 'I looked ... and lo', which has a structural parallel in the sevenfold repetition in Genesis of the phrase: 'and God *saw* ... that it was good' (Gen 1:4, 10, 12, 18, 21, 25). In the last occurrence of this phrase in Genesis 1, the word 'and lo' (והנה) is added: 'And God *saw* ... *and lo* it was very good' (Gen 1:31).

(d) The collocation of the words '... the earth ... and the heavens ...' (Jer 4:23) has similarities with Gen 1:1, in which the same collocation is used, although in reverse order ('In the beginning God created the heavens and the earth').

(e) The word 'bird' occurs in both texts (Jer 4:25b; Gen 1:20).

(f) The expression 'there was no man' (Jer 4:25a) could refer to the sixth day when God created man (Gen 1:27).

(g) The fruitful land that became a desert (Jer 4:26) can be related to Gen 1:11 where the creation of the fruitful land is described on the third day.

I do not intend to exclude completely the possibility that the last two references do not refer to the first creation story in the Bible but rather to the second (Gen 2:4b–25).[16] In the first place there is Jer 4:25a (‘*there was no man*’), which could refer also to Gen 2:5 (‘*there was no man* to till the ground’). Gen 2:4b–6 describes in its own way the situation of chaos before the creation of man. Not only is there no man to till the ground, there is also a rainless land without vegetation, in short, a desert.[17]

In the second place there is Jer 4:26. This verse does not seem to refer to the creation story. It seems more likely that it refers to the settlement in Canaan. However, although nowhere in the Hebrew Bible is כרמל compared explicitly to the Garden of Eden, I would like to point out the possibility that the word ‘fruitful land’ (כרמל) not only refers to Gen 1:11, but possibly also evokes the Garden of Eden. If this is true, one could say that the creation of the Garden of Eden, described in Genesis 2, is annulled. In that case Jer 4:26 returns to a situation before its creation. We can now compare texts in the Hebrew Bible in which the Garden of Eden is transformed into a desert (Joel 2:3: ‘The land is like the garden of Eden before them, but after them a desolate wilderness, and nothing escapes them’) or a desert that is transformed in the Garden of Eden (Isa 51:3: ‘For YHWH will comfort Zion; he will comfort all her waste places, and will make her wilderness like Eden, her desert like the garden of YHWH; joy and gladness will be found in her, thanksgiving and the voice of song’; Ezek 36:35: ‘This land that was desolate has become like the garden of Eden; and the waste and desolate and ruined cities are now inhabited and fortified’).

Finally, Jer 4:24 mentions the mountains and the hills. These do not occur in the description of the creation of heaven and earth in Genesis 1. Nevertheless, in the Hebrew Bible they are connected with the creative acts of YHWH. I refer to Ps 65:7 (‘[Thou] by thy strength hast established the mountains’), to Ps 90:2 (‘Before the mountains were brought forth, or ever thou hadst formed the earth and the world’) and especially to Proverbs 8:25 (‘Before the mountains have been shaped, before the hills, I was brought forth’).

---

16 Cf. Weippert, *Schöpfer*, 52–54.

17 The expression ‘there was no man’ occurs in the same line as the mention of the birds (Jer 4:25). It is possible that the absence of man means that there is no man to have dominion over the birds (cf. Gen 1:26). See Borges de Sousa, ‘Jer 4,23–26’, 423. If this is the case, the reference to Gen 2:5 is less probable.

4. *Final remarks*

The conclusion seems to be inevitable that the prophet has depended on the description of heaven and earth in several traditions with regard to the creation, but especially to the *text* of Genesis 1. Of course, he has modified the text and the traditions related to it in order to express his vision, in which he describes the creation of chaos out of order. He does not seem to be describing a *literary* return to the chaotic primary condition before creation. The comparison is being used as an image. The judgment on Israel is *as* a return to the original chaos.[18]

If the vision does refer to the *text* of Genesis 1, then the text must be later than that. Usually, chapter 4 is interpreted as a *prediction* of the destruction of 586 BC. However, it could be a prediction in the narrated time. This does not mean that the time of the narrator is also before the destruction, it could be afterwards. Because Jer 4:23–26 is well integrated in the book, this could apply to a larger unit than just these verses. If we interpret chapter 4 not as the description of the situation of Judah and Jerusalem before the destruction of 586 but afterwards, it is possible to read Jer 4:23–26 as an interpretation of these events.[19] The author seems to have made use of the story of the creation to interpret the chaos after the destruction. The events should thus be understood as a doom coming from YHWH.[20]

---

[18] Cf. e.g., Lundbom, *Jeremiah 1–20*, 257.

[19] These conclusions coincide with Borges de Sousa, 'Jer 4,23–26', 427.

[20] In this respect, Borges de Sousa points out a last interesting parallel with Genesis 1. Whereas Genesis 1 presents order as the creative power of *Elohim*, the author of Jer 4:23–26 describes the return to chaos as the destructive power of YHWH. Cf. Borges de Sousa, 'Jer 4,23–26', 427.

'LIGHTS SERVING AS SIGNS FOR FESTIVALS' (GENESIS 1:14B) IN ENŪMA ELIŠ AND EARLY JUDAISM

Eibert J.C. Tigchelaar

1. *Introduction*

According to the translation of *The New English Bible* (*NEB*), God said: 'Let there be lights in the vault of heaven to separate day from night, and let them serve as signs both for festivals and for seasons and years. Let them also shine in the vault of heaven to give light on earth' (Gen 1:14–15a). The Hebrew of Gen 1:14b is problematic, both syntactically and semantically, and a verbal rendering results in 'and let them be' (or: 'they shall be') 'for signs and for festivals and for days and years'.[1] The topic of this paper derives from the *NEB* rendering, in particular the concept of the luminaries 'serving as signs for festivals'.[2]

Most modern non-Jewish commentators on Genesis do not focus on the phrases of Gen 1:14b, but limit themselves to general comments on this semi-verse, for example, that the luminaries govern both the cultic and the civil calendar, or that the religious festivals were connected to the calendar. Instead, scholars emphasize entirely different points. Gen 1:14–19 is often read as a denial of the common belief that sun and moon were deities, and perhaps as a reaction to astrology which is based on the principle that the planets and other luminaries influence

---

[1] Thus, e.g., ASV, KJV, and NRSV. The Hebrew runs והיו לאתת ולמועדים ולימים ושנים. The problems involve the meaning of אתת and מועדים, as well as the relation between the three terms preceded by ל, 'for'. E.A. Speiser, *Genesis* (AB 1), Garden City, NY 1964, 6; the JPS; and N.M. Sarna, *Genesis: The Traditional Hebrew Text with the New JPS Translation*, Philadephia 5749/1989, 9 understand לאתת ולמועדים as a hendiadys 'signs for the set times', and ולימים ושנים, 'for the days and the years', as a specification (the *waw* in ולימים is read as an explicative *waw*). Alternatively, one may, with C. Westermann, *Genesis 1–11* (BKAT 1.1), Neukirchen-Vluyn 1974 and many other scholars read 'for the festivals and for the days and years' as a specification of 'to be signs' (the *waw* in ולמועדים is read as an explicative *waw*). Cf., e.g., the Jerusalem Bible: 'and let them indicate festivals, days and years'. Usually מועדים is translated as either 'festivals', 'set times' or 'seasons', but the *NEB* with which this paper started, complicates things by rendering ימים, 'days', by 'seasons'.

[2] Originally, I was asked to deliver a paper on the following subject: 'The Relationship between Creation, Cosmos, and Liturgy: "Lights serving as Signs for Festivals"'.

worldly affairs.[3] Scholars generally highlight the following aspects on the basis of a reading of the text against its assumed religio-historical background:[4] first, Gen 1:16 emphatically tells that God *made* the two great lights and the stars; second, the use of the terms 'greater light' and 'lesser light' in 1:16, instead of 'sun' and 'moon', which words also served as divine names, may indicate that the author wanted to demythologize the luminaries;[5] third, the enumeration of the functions of the luminaries explicitly ignores, and therefore apparently denies, the rule of the luminaries over the fate of earth or mankind.[6]

In this paper, the issue is not the meaning of the entire section Gen 1:14–19, and not even how the Priestly authors of Genesis 1 intended Gen 1:14b to be understood. Nor shall I deal with the large theme of the relation between cosmos, creation, and liturgy. Instead this study is confined to two questions. First, to what extent the concept of luminaries serving as signs for festivals might also be implied in the *Enūma eliš*, a text which is often referred to in the discussion of the creation account of Genesis 1. Second, how Early Jewish texts interpreted, explicitly or implicitly, Gen 1:14b.

2. *Enūma eliš*

Ever since George Smith's 1876 *The Chaldean Account of Genesis*, the so-called Babylonian *Epic of Creation* or *Enūma eliš* has been used to elucidate Genesis 1.[7] Up to the middle of the twentieth century, biblical scholars posited a relation between Genesis 1 and the *Enūma eliš*, and some even thought that the Priestly authors used the *Enūma eliš* as their

---

[3] Westermann, *Genesis 1–11*, 179: the intention of the section 1:14–19 is that 'Sonne und Mond sind Gottes Geschöpfe, darin liegt ihre Würde, darin liegt ihre Grenze'.

[4] G.J. Wenham, *Genesis 1–15* (WBC 1), Waco 1991, 21.

[5] On the sun- and moon-cult, and the names of these deities, cf. E. Lipinski, 'Shemesh', in: *DDD*², 764–768, and B.B. Schmidt, 'Moon', in: *DDD*², 585–593.

[6] W.H. Schmidt, *Die Schöpfungsgeschichte der Priesterschrift* (WMANT 17), Neukirchen-Vluyn 1973³, 110–111, 117–120.

[7] For recent translations of *Enūma eliš* cf. J. Bottéro & S.N. Kramer, *Lorsque les dieux faisaient l'homme: Mythologie mésopotamienne* (Bibliothèque des histoires), Paris 1989, 602–679, with extensive comments; B.R. Foster, *Before the Muses: An Anthology of Akkadian Literature*, 2 vols, Bethesda, MD 1993. Cf. also Foster, *From Distant Days: Myths, Tales, and Poetry of Ancient Mesopotamia*, Bethesda, MD 1995, an edited selection from *Before the Muses*. The most recent comprehensive discussion, with basic bibliography, is H.L.J. Vanstiphout, '*Enūma eliš* as a Systematic Creed: An Essay', *Orientalia Lovaniensia Periodica* 23 (1992) 37–61.

prime literary source.[8] Many of the purported analogies between *Enūma eliš* (*Ee*) and Genesis 1 are of a general nature, but the terminological and conceptual correspondences between Gen 1:14–19 and *Ee* Tablet 5 are more specific.

The first half of *Enūma eliš* describes the births and conflicts of the gods, as well as the battle between Marduk and Tiamat. This theogony and theomachy are in a sense the mythological introduction to the second part, which hails Marduk as organizer and provider. The body of Tiamat serves as the material from which the Lord arranges the cosmos. By dividing Tiamat he forms the firmament and separates the waters. He founds the divine sanctuaries in Esharra, that is heaven, and then proceeds with the instalment of the stars and the calendar (*Ee* 5.1–10), of which I quote the first four lines:[9]

> He made the position(s) for the great gods, (5.1)
> He established (in) constellations the stars, their counterparts.
> He marked the year, described its boundaries,
> He set up twelve months of three stars each.

In *Ee* 5.11–24 Marduk installs the moon:[10]

> In her liver he established the zenith, (11)
> He made the moon appear, entrusted (to him) the night.
> He assigned to him the crown jewel of night time, to mark the days (of the month):
> 'Every month, without ceasing, start off with the (crescent) disk.
> 'At the beginning of the month, waxing over the land, (15)
> 'You shine with horns to mark six days,
> 'At the seventh day, the disk as [ha]lf.
> 'At the fifteenth day, you shall be in opposition, at the midpoint of each [month].
> 'When the sun f[ac]es you from the horizon of heaven
> 'Wane at the same pace and form in reverse. (20)
> 'At the day of di[sappeara]nce, approach the sun's course,

---

[8] The use of *Enūma eliš* in its relation to Genesis 1 is the subject of the dissertation of Joan Heuer Delano, 'The "Exegesis" of *Enuma Elish* and Genesis 1—1875 to 1975: A Study in Interpretation' (diss. Marquette University; Milwaukee, WI 1985).

[9] Translations are, with minor modifications, from Foster, *Before the Muses*. For the text of *Ee* 5, see B. Landsberger & J.V. Kinnier Wilson, 'The Fifth Tablet of *Enūma eliš*', *Journal of Near Eastern Studies* 20 (1961) 154–177, though the two additional sources E and F do not add much to the already known text of 5.1–26 as published by, e.g., R. Labat, *Le Poème babylonien de la création*, Paris 1935.

[10] *Ee* 5.23–46 are incomplete, and it is not entirely clear where the section on the sun starts.

'On the [ ] of the [thirt]ieth day, you shall be in conjunction with the sun a second time.[11]
'I d[efined?] the celestial sign, proceed on its path,
'[ ] approach each other, and render (oracular) judgment.

The section on the sun (*Ee* 5.25–44) is almost entirely lost, apart from some references to year and New Year's Day or *zagmukku*. Tablet 5 continues with the regulation of the weather and the earth, the distribution of Marduk's trophies, the celebration of Marduk as king, and Marduk's promise of an earthly abode for the gods, *Esagila* in the city of Babylon.

In *Enūma eliš*, the actual description of the creation or organization runs only from 4.135–5.66, of which the largest part is taken up by the instalment of the luminaries. The description of the moon focuses on the calendrical aspects: the moon marks by means of its regular waxing and waning the days of the month (5.12–22). The preserved words in 5.23–24 suggest that these lines refer to the 'sign' function or astrological function of the moon.[12] In short, the moon is described here as a marker of time, and as a provider of signs or omens.

In the neo-Babylonian period, *Enūma eliš* was recited on the fourth day of the Akitu or New Year's Festival. In the past, scholars posited a close relationship between myth and ritual, and launched the hypothesis that scenes from *Enūma eliš* were re-enacted during the Akitu-festival.[13] Even though there is no evidence for a dramatic performance of the *Enūma eliš*, the text was used, at least in later times, in ritual. Other elements of the text, such as the focus on the foundation of sanctuaries, especially the Esagila in Babylon, and the revelation of the fifty names of the Lord which should be invoked, are also somehow related to ritual.

---

[11] H.L.J. Vanstiphout, '*Enūma eliš*: Tablet V Lines 15–22', *Journal of Cuneiform Studies* 33 (1981) 196–198 proposes a different restoration and translates: 'So that the next fifteenth day('s moon) will be again equal to the sun'. In that case *Ee* 5.15–22 describes a one and one-half lunar cycle.

[12] J. Bottéro & S.N. Kramer, *Lorsque les dieux faisaient l'homme: Mythologie mésopotamienne* (Bibliothèque des histoires), Paris 1989, 664.

[13] Labat, *Le Poème babylonien*, 59–63. Because of the repetition of this hypothesis in many general works, such as, e.g., G. Widengren, *Religionsphänomenologie*, Berlin 1969, 151, the idea of a dramatic reenactment of the *Enūma eliš* has proved to be persistent up to present times. Cf., for criticisms: K. van der Toorn, 'The Babylonian New Year Festival: New Insights from the Cuneiform Texts and Their Bearing on Old Testament Study', in: J.A. Emerton (ed.), *Congress Volume Leuven 1989*, Leiden 1991, 331–344; B. Sommer's review of Baltzer's commentary on Deutero-Isaiah in *Review of Biblical Literature* 02/2003.

A ritual meaning of the cosmological materials of *Ee* 5 is less apparent, but nonetheless some connections between cosmology and ritual have been observed. Landsberger and Kinnier Wilson emphasize that the terms 'position' (*manzāzu*) and 'counterpart' (*lumāšu*) in *Ee* 5.1–8 should not primarily be interpreted as astronomical terms, but that *manzāzu* refers to the image-stands of deities or statues and that the basic meaning of *lumāšu* is 'twin-image', or 'replica': heaven with its stars in their positions, is described as the heavenly sanctuary of the great gods, with image-stands for their replicas.[14] Livingstone argues more explicitly for a ritual importance of the instalment of the moon on the basis of a mathematical tablet (K2164+2195+3510) of the Neo-Assyrian scribe Nabûzuqupkēna.[15] K2164+ obv. 3 quotes *Ee* 5.17: 'At the seventh day, the disk as [ha]lf', then continues with a description of the 14th, 15th, 21st, 27th and 28th day, and line 24 closes this section with the explicit quotation of *Ee* 5.21: 'At the day of disappearance, approach the sun's course'. The days mentioned in *Enūma eliš* and in the Commentary are not entirely identical, but the correspondence would be the practical importance which these days had in the ritual calendar. Thus, *Atrahasis* refers to these same days, the first, seventh and fifteenth, as days of ritual purification (*Atrahasis* 1.203–204 and 215–216).[16] Livingstone argues that, according to the Neo-Assyrian commentary, the moon's shape is an indication of the day (in the month). The moon should follow her course in order that the rituals will take place on the right day. This commentary would then provide us with a Mesopotamian example of a luminary serving as a sign for ritual days. However, against Livingstone it should be maintained that whereas a ritual significance of these specific days may be possible,[17] neither *Enūma eliš*, nor the explanatory work K2164+, explicitly refers to this aspect.

On the other hand, the elaboration of K2164+ obv. 20–23 on the astro-theological meaning of the 27th day, concurs with other texts

[14] Landsberger & Kinnier Wilson, 'Fifth Tablet', 170–171: 'These terms … are thus considered properly at home when used in connection with the temple'. For a recent discussion of the cosmological elements of *Enūma eliš* cf. W. Horowitz, *Mesopotamian Cosmic Geography* (Mesopotamian Civilizations 8), Winona Lake, IN 1998, 107–129.

[15] A. Livingstone, *Mystical and Mythological Explanatory Works of Assyrian and Babylonian Scholars*, Oxford 1986, 22–29 (text), 38–44 (commentary).

[16] 1.201–206: 'Enki made ready to speak, / And said to the great gods, / "On the first, seventh, and fifteenth days of the month, / Let me establish a purification, a bath. / Let the one god be slaughtered, / Then let the gods be cleansed by immersion"'.

[17] But cf. Mark Cohen, *The Cultic Calendars of the Ancient Near East*, Bethesda, MD 1993, who demonstrates the variety of the cultic calendars throughout Mesopotamia.

which stress the astrological aspects of the creation of the luminaries. K2164+ obv. 22–23 state that Shamash and Sin 'make the decrees concerning the land' (*purussê māti iparrasū*) and 'give the signs for the land' (obv. 23 *saddu ana māti inaddinū*), a statement which, according to Livingstone, apparently reflects the conviction that fates were determined on the day of the disappearance of the moon.[18] Yet, the 27th day is close to, but not identical to the day of the disappearance of the moon, and the number 27 seems to be based on the arithmetics of the numerical values related to Sin and Shamash. The defining of celestial signs and the making of decrees are also mentioned in one of the prologues to *Enūma Anu Enlil* that is close to the opening lines of *Ee* 5 with regard to terminology and content.[19] The exact meaning of these celestial signs (*giskimmu* or *ittu*) is not specified in these texts, but they seem to refer to the astronomical properties of the celestial bodies. The context of *Ee* 5.23 suggests that GISKIM, 'celestial sign', there refers to the particular shape of the moon, or rather its position in relation to the sun. This specific element in *Ee* 5 raises the question to what extent these celestial signs are related to the אתת, 'signs' of Gen 1:14. It is unclear whether there is an etymological relation between *ittu* and אות, but there seems to be a considerable semantic overlap.[20]

In conclusion: the instalment of the luminaries in *Ee* 5 (and even more so in the Prologue to *Enūma Anu Enlil*) is explicitly connected to calendrical, astronomical and astrological matters. The 'signs' in the phrase *(w)adû(m) giskimma*, 'determine the signs', may be compared to אתת in Gen 1:14. Yet, in the preserved parts of *Ee* 5 there is no explicit reference to the luminaries in relation to the festivals, or to the cultic calendar in general. It may be the case that the days mentioned in *Ee* 5 were commented upon by Nabûzuqupkēna because of their importance in the ritual calendar, but his explanatory tablet does not explicitly refer to the cult, whereas it does discuss the astrological functions of the sun and moon.

[18] Livingstone, *Mystical and Mythological Explanatory Works*, 41–42.

[19] Cf. text and discussion in Landsberger & Kinnier Wilson, 'Fifth Tablet', 172, and, most recently, Horowitz, *Mesopotamian Cosmic Geography*, 146–147.

[20] *AHw* 1:405–406, distinguishes between **ittu(m) I**, **idatu**, 'das Besondere', which von Soden tentatively connects to the root *w'd* (cf. מועד!) and **ittu(m) II**, 'Zeichen', which he connects to Hebrew אות, although he acknowledges that *ittu* I and II are 'nicht immer sicher unterscheidbar'. CAD does not differentiate between these two words.

3. *Ben Sira*

In the book of Sirach, dated in the first quarter of the second century BCE, there are three sections which use the term **מועד** in connection with luminaries. Sir 43:6–8 and 50:6–7 are two key-passages which are usually discussed with regard to calendrical matters.[21] As part of the praise of God's creation, the hymn in Sirach 43 extols the sun, moon, and stars, and ascribes to the moon the regulation of the seasons, the times, and the festal days. The Masada manuscript, restored on the basis of MS B, reads as follows:[22]

וגם [י]רח יאריח עתות
ממ[שלת קץ ואות עולם]
לו מו[ע]ד וממנו חג

And the moon, too, marks the times,
a rule for the fixed time and an eternal sign
To it belongs the festival, and from it (derives) the feast.

The combination of **אות**, 'sign', **ממשלה**, 'dominion', 'rule' (Sir 43:6) and **מועד** (43:7) suggests an indebtedness to Gen 1:14 and 16. These three hemistichs relate the moon's dominion to four different calendrical terms (**עתות**, 'times'; **קץ**, 'fixed time'; **מועד**, 'festival', 'season'; **חג**, 'feast'), whereas in the previous verses 43:2–5 the sun is associated with light and heat, but not with calendrical matters.

In the hymn to Simon, the high-priest, in Sirach 50, his coming out of the holy of holies is described as the emergence of a star, the moon, the sun, and the rainbow. Sir 50:6 compares it to 'the full moon (**ירח מלא**) on festival days (**בימי מועד**)'. These two passages may be taken as evidence for a luni-solar cult calendar in the days of Sirach, whereas

[21] Cf. B.G. Wright, '"Fear the Lord and Honour the Priest": Ben Sira as Defender of the Jerusalem Priesthood', in: P.C. Beentjes (ed.), *The Book of Ben Sira in Modern Research*, Berlin & New York 1997, 189–222 at 204–208; J.C. VanderKam, *Calendars in the Dead Sea Scrolls: Measuring Time*, London 1998, 27.

[22] The LXX has Καὶ ἡ σελήνη ἐν πᾶσιν εἰς καιρὸν αὐτῆς, ἀνάδειξιν χρόνων καὶ σημεῖον αἰῶνος· [7] ἀπὸ σελήνης σημεῖον ἑορτῆς. On this text see A. Minissale, *La versione greca del Siracide: Confronto con il testo ebraico alla luce dell'attività midrascica e del metodo targumico* (AnBib 133), Rome 1995, 211. The Geniza MS B has a different reading, including 43:7a **בם מועד וזמני חוק** with the marginal reading **בו מו ׳ וממנו**. O. Mulder, *Simon the High Priest in Sirach 50: An Exegetical Study of the Significance of Simon the High Priest as Climax to the Praise of the Fathers in Ben Sira's Concept of the History of Israel* (SJSJ 78), Leiden 2003, 128–129 favours the reading in 43:7 of MS B **בם**, 'through them (sc. sun and moon)', against LXX, MS M and the marginal reading of MS B on the grounds that MS B has the *lectio difficilior*!

some scholars detect a propagandist or polemic undertone in these descriptions.[23] Initially, VanderKam played down the evidence of these verses, and argued that in Sirach's days the 364 calendar was the official cultic calendar,[24] but more recently he conceded that it 'is safe to say on the basis of these references that the moon, not the sun, plays the decisive role in connection with the holidays for the author'.[25]

To these two key-passages that indicate a lunar festival calendar one may perhaps add a third one, Sir 33:7–9:

> [7]Why is one day more important than another,
> when (all) the daylight in the year is from the sun?[26]
> [8]By the Lord's wisdom they were distinguished
> and he appointed the different festivals and feasts.
> [9]Some days he exalted and hallowed,
> and some he made ordinary days (NRSV).

These verses deny calendrical importance of the sun with regard to festival days: not the sun determines whether some days are different from other ones, but, through his wisdom the Lord has separated festivals and feasts from ordinary days.

What do these key-passages mean for Ben Sira's reception of Gen 1:14? The only passage which alludes to Gen 1:14–16 is Sir 43:6–7. Here, חג, 'feast', is used parallel to מועד, which indicates that Ben Sira understood מועדים in Gen 1:14 as 'festivals', not as 'seasons'. In addition, the separate clause in Sir 43:6bβ ואות עולם, 'and an eternal sign', suggests that Ben Sira did not understand לאתת ולמועדים as a hendiadys 'signs for the festivals', but that the moon itself was understood as an אות, a 'sign'. In these passages, אות seems to refer to celestial phenomena which serve as signs.

It is clear that Ben Sira's statements reflect a lunar calendar. The question remains whether the allusion to Gen 1:14 in the section on the moon in 43:6–7, in contrast to the focus on other aspects in the description of the sun in 43:2–5, implies that Ben Sira deliberately ignored parts of Gen 1:14. This depends on the way one reads the text

---

[23] Most recently Wright, 'Fear the Lord'; for older studies arguing for a polemic against the solar calendar, cf. Minissale, *La versione greca del Siracide*, 211 note 79.

[24] J.C. VanderKam, 'The Origin, Character, and Early History of the 364-Day Calendar: A Reassessment of Jaubert's Hypothesis', *Catholic Biblical Quarterly* 41 (1979) 390–411.

[25] VanderKam, *Calendars in the Dead Sea Scrolls*, 27.

[26] Following the Greek, and reading שָׁנה, 'year', in MS E (corrected reading) rather than שֹׁנה, 'changed' or שׁוֹנה, 'changing' (MS F and the first hand of MS E).

at large. The long section Sir 42:15–43:33 is a hymn on the 'Works of God', or rather, a glorification of the creator. The hymn describes the wonders of creation, especially those of a celestial and meteorological nature. After the introduction, the hymn describes in stanza's of more or less the same length the marvelty of the sun (43:2–5—five lines), the moon (43:6–8—four lines), and the stars and rain-bow (43:9–12—together four lines). In the depiction of the sun, the hymnic focus is on its wonderful heat and light, whereas in the description of the moon its calendrical importance is the theme. From a poetic or hymnic point of view, there is no need to expect explicit reference to calendrical matters in the section on the sun. Ben Sira is writing a hymn, not a commentary on Genesis 1, nor a pamphlet.

Wright, however, argues for an intertextual study of works that are contemporary, and reads Sirach in the light of *1 Enoch* and *Aramaic Levi*. Ben Sira's statements on the celestial bodies and the festivals then serve as a polemic against the solar year found in those compositions. This then would imply a deliberate attempt on Ben Sira's part to ignore the role of the sun in Gen 1:14–19.[27]

4. *Jubilees*

The mid-second century BCE *Book of Jubilees* presents a re-writing of the Book of Genesis.[28] The account of the creation of the luminaries on the fourth day runs as follows in *Jub.* 2:8–10:

> [8]On the fourth day the Lord made the sun, the moon, and the stars. He placed them in the firmament of the heavens to shed light over the whole earth, to rule over the day and the night, and to separate between light and darkness. [9]He appointed the sun as a great sign above the earth for days, for sabbaths, for months, for festivals, for years, for the weeks of years, for jubilees, and for all the cycles of the years. [10]It separates between the light and the darkness and serves for healing so that everything that sprouts and grows on the earth may be well. These three types he made on the fourth day.[29]

[27] Wright, 'Fear the Lord', 206–207 and 221. On the 364-day calendar in *Aramaic Levi Document*, cf. also M.E. Stone, 'Aramaic Levi in Its Contexts', *Jewish Studies Quarterly* 9 (2002) 307–326 at 322–323.

[28] J.C. VanderKam, *The Book of Jubilees* (Guides to Apocrypha and Pseudepigrapha), Sheffield 2001, 17–21 discusses the approximate date of composition and suggests the period between 160–150 BCE.

[29] Translation from J.C. VanderKam & J.T. Milik, *DJD* 13, 17. This is a translation

Van Ruiten shows that *Jubilees* abbreviates the biblical text, for example by removing repetitions, but that the Genesis clause on the 'sign-function' of the luminaries has been drastically changed.[30] Here, two differences are of interest. First, whereas in Gen 1:14 the luminaries, without specification, are said to be signs, *Jub.* 2:9 highlights the sun, which is referred to as a 'great sign', and does not bring up the other luminaries at all. Second, Gen 1:14 mentions 'festivals', 'days and years', but *Jub.* 2:9 expands this list to eight items, namely 'days, sabbaths, months, festivals, years, weeks of years, jubilees, and all the cycles of the years'.

Most items are only preserved in Ethiopic, and have to be retroverted into Hebrew. In both languages some of these words are ambiguous.[31] The second item of the calendrical units is *sanbat* (a loanword derived from Hebrew שַׁבָּת) which may be used both for the 'Sabbath', and for the 'week', even though Ethiopic has a separate word for week (*sabuʿa*). In fact the Hebrew plural שבתות also refers to periods in between Sabbaths, hence to weeks. Thus, most translations render the sixth item *sanbatāta ʿāmatāt* with 'sabbaths of years'. This is misleading, since the word clearly refers to periods of seven years, just like a week is a period of seven days. This not only goes for the construction *sanbatāt* of years, but also for seven-day periods. In short, the second item *sanbatāt* may also mean 'weeks'.[32]

The fourth item, *baʿālāt*, 'festivals', is also ambiguous. In the Ethiopic text of *Jubilees*, *baʿāl* generally corresponds to מועד, 'season', 'fixed time', or 'festival'. In *Jub.* 6:23–29 it is clear that there are four memorial festivals in the year, and 6:29 states explicitly that each of them consists of thirteen weeks. If מועד is indeed 'season', then one has here a list of all the time-units the author uses, namely 'days', 'weeks', 'months', 'seasons', 'years', 'seven-year periods', 'jubilees' (i.e. forty-nine year periods).[33] The last item remains partly in Hebrew, and probably read כל

of the restored lines of *4Q216* (*4QJub*[a]) 6:5–10.

[30] J.T.A.G.M. van Ruiten, *Primeval History Interpreted: The Rewriting of Genesis 1–11 in the Book of Jubilees* (SJSJ 66), Leiden 2000, 35–40.

[31] Cf. the following correspondences: יום = *ʿelat*; שבוע / שבת = *sanbat*; חדש = *warh*; מועד = *baʿāl*; שנה = *ʿāmat*; שבועי שנים = *sanbata ʿāmatāt*; יובל = *ʾiyobēl*; תקופה = *gizē*.

[32] According to the editors *4Q216* 6 reads ול[ש]בתות. It is not possible to verify this reading from the available photographs, and ול[ש]בועות does not seem impossible to me.

[33] A reference to 'weeks of jubilees' (i.e. possibly periods of seven jubilees), is only found in a notice (4:18) on the books of Enoch, not elsewhere in the book of *Jubilees*.

תק]ופות השנים] (*gizē la'āmatāt*).[34] This is not a longer period of time than the preceding 'jubilees', but a summary of all preceding items: all the cycles (or: periods) of years.

The declaration that the sun is placed as a great sign for all these time-units apparently denies any role at all for the moon in calendrical matters. One should note that the list also contains 'months', but that nowhere in the composition the duration of a month is described. If we read the enumeration of items as referring to units of time, than the text states primarily that all calendrical matters are dependent on the sun.

In *Jub.* 6:37 (and in the calendrical section *Jub.* 6:23–38 as a whole) the calendrical matter is related to ritual or liturgy:

> Years will come about for them when they will disturb (the year) and make a day of testimony something worthless and a profane day a festival. Everyone will join together both holy days with the profane and the profane day with the holy day, for they will err regarding the months, the weeks, the festivals, and the jubilee.[35]

This section focuses on the special memorial festival days at the beginnings of the seasons, on the first days of the first, fourth, seventh, and tenth months. That is, the spring equinox, summer solstice, autumn equinox, and winter solstice, which mark the four seasons. The 'error regarding the months' is the error of correlating months to the moon. If the first days of the seasons coincide with the first days of the first, fourth, seventh and tenth month, then the sun, not the moon determines the beginnings of the months. In sum, *Jubilees* deliberately applies the 'sign'-function of Gen 1:14 only to the sun, which not only is the sole determinator of the calendar, but, as a consequence, also serves as the great sign for the festivals.

5. *Rule of the Community*

Many texts from Qumran are related to liturgy, and it would be impossible to discuss all texts relating to 'festivals'.[36] One of these texts is the so-called Calendrical Hymn or Maskil's Hymn of *The Rule of the Commu-*

---

[34] Suggested by VanderKam and Milik in their reconstruction of *4Q216* 6:8. Alternatively reconstruct כל תק]ופות שנים]. The phrase תקופות שנים is attested in *1QM* 10:15.

[35] Similar statements about the erring or forgetting of the first of the month, the sabbath, and the festivals are found in *Jub.* 1:14, and 6:34, 38.

[36] On the term מועד, cf. G. Brin, *The Concept of Time in the Bible and the Dead Sea Scrolls* (STDJ 39), Leiden 2001, 253–263: chap. 19 'The Term מועד in the Scrolls'.

nity (*1QS* 9:26–11). The introduction states 'and with the offering of the lips he shall bless Him at the times which God has decreed' (*1QS* 9:26–10:1) and 10:6 'I will bless Him with the offering of the lips according to the precept engraved for ever'. Taken together, these statements claim that the times of prayer have been determined for ever, an assertion which is also made in *Jubilees* with regard to some of the festivals.

The times of prayer are arranged from daily prayers at the beginning of the hymn to the blessings at the so-called מועדי דרור, 'the appointed-times of liberty', that is, the jubilees, at its very end. The structure of the hymn is not always self-evident, and there are some uncertainties about the beginnings or ends of clauses.[37] The times of the daily prayer are directly related to the daily regular change of light and darkness. This section of the daily cycle of prayer seems to consist of three parallel verses, each referring to morning and evening. The section begins with a reference to dawn, which suggests that for the Qumran community day began in the morning, with the rising of the sun, instead of in the evening with the rising of the moon.[38]

1QS 10:3 'At the entrance of seasons on the days of the new moon' opens the section of the annual cycle. This section deals first with the four 'days of remembrance' which are mentioned in *Jub.* 6:23, namely the first days of the first, fourth, seventh, and tenth month. These four beginnings of the seasons are referred to three times in the hymn in *1QS* 10:3–7, with an interruption in line 5 which refers to the beginning of months and holy days. This annual cycle section repeatedly has the word מועד, which should be rendered 'seasons' in some cases, but 'appointed times', or the like, in other cases. Within the *1QS* collection, this use of מועד is already introduced in the beginning of the Rule, which prescribes that any one who enters the covenant shall not stray from any of God's orders concerning their times: 'they are not to advance their periods, nor to postpone any of their מועדים' (*1QS* 1:14–15). These instructions only make sense in a situation where one has an irregular luni-solar calendar in contrast to a fixed and regular solar one.

The scant reference to the common Jewish feasts and holy days, as opposed to the long and repeated description of the prayers on the

[37] See P.S. Alexander and G. Vermes, *DJD* 26, 120, for a neatly-arranged table of times of prayer of this hymn, distinguishing a daily, an annual, and a septennial cycle.

[38] Cf. also *4Q408* 3+3a 8–10 where the mention of the creation of 'the morning' precedes the mention of the creation of 'the evening'.

seasonal days, serves to accentuate the differences between the author's group and other groups. It does not mean to say that the Sabbath or, for example, the Feast of Weeks were less important in the Qumran community. Apparently, *1QS* has the same calendar as *Jubilees*, even though its terminology is somewhat different. The sun determines the calendar, which has been decreed for all times, and in each period.

This hymn does not directly refer to Gen 1:14, but it shares, like most Qumran texts, the calendar of *Jubilees*. Next to the frequent mention of **מועד**, the word **אות**, 'sign', is also used, namely in a side-remark in *1QS* 10:4, as a comment on the four first days of the seasons:

> When they (i.e., the seasons) are renewed, it is a great day for the holy of holies, and a *sign* for the release of eternal mercies.

Vermes and Alexander suggest that we have here a distant echo of the Noah-covenant (Gen 9:12), which refers to the rainbow as a 'sign of the covenant'.[39] Below, in the section on the semantics of **אות**, I will return to this suggestion.

6. *4QInstruction*

Sirach and Jubilees relate the festivals to one of the luminaries, and use the terms **אות** and **מועד** which are found in Gen 1:14. A fragmentary, sapiential work (*4QInstruction*) from Qumran, seems to interpret Gen 1:14 in an astrological manner.[40] The first column of the composition has been preserved in *4Q416* 1. I identified some fragments of other manuscripts which seem to overlap, and reconstructed the first lines of this fragment as follows:[41]

---

[39] *DJD* 26, 123.

[40] The different manuscripts of this composition are published by J. Strugnell & D.J. Harrington (*4Q415–4Q418c* and *1Q26*) and T. Elgvin (*4Q423*) in *DJD* 34.

[41] E.J.C. Tigchelaar, *To Increase Learning for the Understanding Ones: Reading and Reconstructing the Fragmentary Early Jewish Sapiential Text 4QInstruction* (STDJ 44), Leiden 2001, 175–176. In my 'Towards a Reconstruction of the Beginning of 4QInstruction (4Q416 Fragment 1 and Parallels)', in: C. Hempel, A. Lange, and H. Lichtenberger (eds), *The Wisdom Texts from Qumran and the Development of Sapiential Thought* (BETL 149), Louvain 2002, 99–126, I dealt with the material and technical aspects of this reconstruction in extenso.

1	כל רוח]	כוכבי אור]
2	ולתכן חפצי [	ירוצו מעת עולם]
3	מועד במועד ו[	ואין להדמות בכושר ילכו]
4	לפי צבאם למשׁ[ור במשורה ול	ת לממלכה]
5	וממלכה למד[ינה ומדינה לאיש ואיש	
6	לפי מחסור צבאם [ומשפט כולם לו	
7	וצבא השמים הכין ע[ל	ומאורות]
8	למופתיהמה ואתות מו[עדיהמה	ו[
9	זה לזה וכל פקודתמה י[שלימו ו]ספרו[	

1. every spirit [... stars of light,]
2. and to allot the tasks of [... they run from eternal time,]
3. season upon season, and [... without standing still. Properly they go,]
4. according to their host, to ke[ep station (?), and to ... for kingdom]
5. and kingdom, for pr[ovince and province, for each and every man,
6. according to the poverty (?) of their host. [And the regulation of them all belongs to Him
7. And the host of heavens He has established ov[er ... and luminaries]
8. for their portents, and the signs of [their] se[asons
9. one after another. And all their assignments [they] shall [complete, and they shall] count (?) [

In an earlier discussion of this section, I presented a series of short line by line comments, and concluded with an overall summary:

> In short, this unit consists of a description of the orderly courses of nature in accordance with God's commands ... The preserved part of the text deals with the tasks of the luminaries, i.e., following their proper courses at the proper times (perhaps the text used the expression בכושר), in order that men may know the set times of the festivals (cf. the terminology in line 8).[42]

However, is this really the point of this section? The reconstructed text of *4Q416* 1:7–8 reads ומאורות למופתיהמה ואתות מו[עדיהמה, 'and luminaries for their portents, and the signs of [their] se[asons'. Three of these terms, מְאֹרֹת, 'lights',**אֹתֹת**, 'signs', and מוֹעֲדִים, 'seasons', are found in Gen 1:14, whereas the combination of אתות, 'signs', and מופתים, 'portents', is very usual in the Hebrew Bible.[43] But does *Instruction* discuss the proper courses of the luminaries, 'in order that men may know the set times of the festivals'? The many preserved fragments of the com-

[42] Tigchelaar, *To Increase Learning*, 179.

[43] For the juxtaposition of the plural 'signs and portents', cf. Exod 7:3; Deut 4:34; 6:22; 7:19; 13:2, 3; 26:8; 28:46; 29:2; 34:11; Is 8:18; 20:3; Jer 32:20, 21; Ps 78:43; 105:27; 135:9; Neh 9:10. The singular 'sign and/or portent' is found in Deut 13:2, 3; 28:46; Is 20:3.

position do not display any interest in cultic or calendrical matters. The quoted text has, on the other hand, several clues which suggest that the authors allowed for some kind of astrology.

The first commentators of *Instruction* recognized the deterministic, even predestinatory mood of the text,[44] and some referred to the term בית מולדים, literally 'house of origins', or perhaps 'house of birth', which may be compared to Syriac *bet yalda* which means, in Bardaisan, 'horoscope'.[45] If one reads the introduction of *Instruction* in this light, one arrives at a different understanding. The issue is then not about the calendrical or liturgical aspect of the courses of the luminaries, but about the fact that they disclose, by means of portents and signs, what is going to happen. In this reading the clause in lines 4–5, 'for kingdom and kingdom, for province and province, for each and every man', makes more sense.[46] The point made here is that the luminaries disclose both the fate of kingdoms and provinces, as was recognized already in Babylonian times, and the fate of individuals. This precisely, was the turning point in astrology in the early Hellenistic period, namely the development of personal horoscopes. The text remains utterly fragmentary, though some of the well preserved fragments corroborate the possibility of an astrological dimension. *4Q416* 1:7–8 has the partially preserved statements that 'God established the host of heaven over' something, and 'the luminaries to be their portents and signs of' (perhaps) their seasons. It is not known to whom or what the suffix in 'their portents' refers, and whether one should reconstruct 'their seasons'. It is clear, however, that by correlating 'portents' to 'signs', the word 'signs' receives a different meaning, namely of 'signs of coming events'. This also is the interpretation of Philo of Gen 1:14 in his *De opificio mundi* 58:

> They have come into existence, as he himself said, not only for the purpose of sending forth light, but also in order to reveal in advance signs

---

[44] A. Lange, *Weisheit und Prädestination: Weisheitliche Urordnung und Prädestination in den Textfunden von Qumran* (STDJ 18), Leiden 1995, 46–92.

[45] See, in particular, M. Morgenstern, 'The Meaning of בית מולדים in the Qumran Wisdom Texts', *Journal of Jewish Studies* 51 (2000) 141–144.

[46] T. Elgvin, 'Qumran and the Roots of the Rosh Hashanah Liturgy', in: E.G. Chazon (ed.), *Liturgical Perspectives: Prayer and Poetry in Light of the Dead Sea Scrolls* (STDJ 48), Leiden 2003, 49–67 at 52–53, also discusses *4Q416* 1, but interprets the entire section as refering to judgment. He reads the clause in lines 4–5 on the kingdoms and provinces in the light of the Rosh Hashanah liturgy according to which God determines on Rosh Hashanah the fates of the countries for the coming year. However, the issue there is not judgment, but the determination of fate. Elgvin himself refers (p. 61) to *Jub.* 12:16–18, for which see below.

> of future events. By observing their risings or settings, or eclipses, or their appearances or occultations, or other variations in their movements, humans make predictions about what will happen, the supply or lack of crops ... alterations in the annual seasons, whether in the form of a wintry summer or a scorching winter ...[47]

One may compare this to the passage in *Jubilees* on the observance of the stars (*Jub.* 12:16–17):

> [16]And in the sixth week, in the fifth year thereof, Abram sat up throughout the night on the new moon of the seventh month to observe the stars from the evening to the morning, in order to see what would be the character of the year with regard to the rains, and he was alone as he sat and observed. [17]And a word came into his heart and he said: 'All the signs of the stars, and the signs of the moon and of the sun are all in the hand of the Lord. Why do I search (them) out?'

In this light, the 'signs of the seasons' of *4Q416* 1:8 should be interpreted as those 'signs of the luminaries' that indicate the characters or fates of the coming seasons.

7. *The Semantics of* אות *and Gen 1:14*

In his introduction to *4Q319* (*4QOtot*), Ben-Dov discusses the meaning of אות, especially with regard to its use in *4QOtot*, and refers to it as a 'polyvalent cosmological term'.[48] The presence of the word in Gen 1:14 triggered its use in this calendrical text, but the author of *4QOtot* combined the meaning of אות in Jer 10:2, 'a celestial phenomenon that has significance on the earthly plane', with the meaning 'standard', as found in the military context of the *War Scroll*:

> the observable heavenly phenomena serve as standards for the priestly courses of Gamul and Shekaniah, with each sign figuratively 'leading' the next three years until the appearance of the following sign.

In *4QOtot*, this sign is 'the specific relation of the sun and moon at the time of Creation, which is repeated after three years', that is, either full

---

[47] Translation adapted from D.T. Runia, *On the Creation of the Cosmos according to Moses: Introduction, Translation and Commentary* (Philo of Alexandria Commentary Series 1), Leiden 2001, 60. For comments on the section see 205–206.

[48] *DJD* 21, 208–210 at 209.

moon, or rather new moon, at the vernal equinox.[49] This very specific technical use of אות in *4QOtot* is not found in the texts discussed above.

In the context of the description of the courses of the luminaries, *4QInstruction* uses אות in the meaning of a sign of a coming event. This concurs closely with Philo's description and the passage in *Jubilees*. Such meaning of אות stands close to the Mesopotamian idea of the luminaries giving the (yearly?) signs, that is, determining the fates, for the lands. According to this interpretation, לְאֹתֹת in Gen 1:14b should not be taken together with the following words, but refers to one of the tasks of the luminaries: they have been created to serve as signs of future events.[50]

In *Jub.* 2:9 the sun is placed as a great sign (*la-te'emert 'abiyy*) above the earth, for days, weeks, etc. The phrase is a conflation of two elements of Gen 1:14–17: a 'great light' and 'sign'. Apparently the author of *Jubilees* understood the syntax of Gen 1:14 like many modern commentators: 'and' in 'and for seasons/festivals' is explicative.[51] The statement that the sun is a 'sign' with regard to the calendrical periods means that the sun, by its movements and varying positions marks the different units of the calendar. In this sense, the sun itself may be referred to as a sign. A similar use is implied in Sir 43:6 which calls the moon 'an eternal sign'.

Both terms, 'a great sign' and 'an eternal sign', are also used for the Sabbath. In *Jub.* 2:1, 17, and 21 the Sabbath is called a 'sign'; in 2:17 *Jubilees* uses the phrase 'a great sign', the same phrase it just used in 2:9 with regard to the sun. It is likely that this was influenced by Exod 31:12–17, which describes the sabbath as a sign, and in 31:17 as an eternal sign. This use of אות with regard to the Sabbath, may lie behind the use of אות in *1QS* 10:4: 'a sign of the release of the everlasting mercies'. The semantics have been extended from the Sabbath as a day which is a sign (that the Lord sanctifies the Israelites), to the מועדים, 'the seasonal festivals', which are now referred to as 'great days' and 'signs of the release of the everlasting mercies'.

---

49 *DJD* 21, 209–210.

50 As stated explicitly by Philo, *De opificio mundi* 55: 'These he established in heaven ... firstly to give light, secondly for signs, then to give the right times for the annual seasons, and finally for days and months and years'. Since Philo read σημεῖα in the LXX, his interpretation may be more influenced by Plato (*Timaeus* 40C9-D2) than by an understanding of אות. Cf. Runia, *Creation of the Cosmos*, 205.

51 Not less than three Genesis manuscripts from Qumran confirm the reading of the MT of ולמועדים with *waw* (*4QGen*^{b} 1 i 17; *4QGen*^{g} 2:3; *4QGen*^{k} 2:3).

8. *Conclusion*

The reader of the creation account of Genesis 1 may interpret the concise statement of 1:14a in different manners, due to syntactic ambiguities and the polyvalence of the words אתת and מוֹעֲדִים. The description of the instalment of the moon in *Enūma eliš* focuses on its function as marker of time and provider of signs and omens. Neither *Enūma eliš*, nor the commentary K2164+ explicitly connect the moon to cultic matters or ritual. A connection between Gen 1:14, calendar and ritual is present in three Early Jewish texts (Sirach, *Jubilees*, *1QS*), be it that they differ with regard to the question which luminary serves as a sign for the festivals. A different reading of the verse is implied in *Instruction*. This text is not concerned with either calendar or ritual, and regards the luminaries as providers of signs for coming events.[52]

[52] For comments and information I would like to thank George van Kooten and Herman Vanstiphout.

CREATION IN THE DEAD SEA SCROLLS

Florentino García Martínez

Introduction

The topic of creation in the Dead Sea Scrolls can be approached from many different perspectives. A few years ago, we concentrated on the 'micro level,' on the two parallel accounts of the creation of man and woman in Genesis and on its interpretation.[1] The idea of this year's meeting was to focus on the 'macro creation,' and my task was to examine how the 'creation' on this level has been interpreted in the Scrolls. To me this implies a certain level of abstraction, of going a step further than the narrative of the biblical text of Genesis in which God's creative action is described using the verb ברא, but where we do not find an abstract name to designate the divine action or all things created. In fact, this level of abstraction, if we may judge from the absence of a name for the results of God's creative act in a general way or this action in itself, is absent from the entire Hebrew Bible, with the exception perhaps of Num 16:30, which is a notoriously problematic verse.

1. *The abstract substantive for 'creation' in the biblical texts*

Within the discourse of Moses which precedes the punishment of Dathan and Abiram, the MT puts the following words into the mouth of Moses:[2]

> [I]f these men die as all men do, if their lot be the common fate of all mankind, it was not the Lord who sent me to do all these things; ואם בריאה יברא יהוה,[3] so that the ground opens its mouth and swallows them

---

[1] G.P. Luttikhuizen (ed.), *The Creation of Man and Woman: Interpretation of Biblical Narratives in Jewish and Christian traditions* (TBN 3), Leiden 2000.

[2] According to the JPS translation.

[3] 'But if the Lord brings about something unheard of' according to the JPS translation.

> up with all that belongs to them, and they go down alive into Sheol, you shall know that these men have spurned the Lord (Num 16:29–30).

The problematic phrase, ואם בריאה יברא יהוה, can be literally translated: 'If the Lord will create a creation,' if we ascribe בריאה the meaning the word has in later rabbinic Hebrew, where the basic word to designate 'creation' has the abstraction level we are looking for.[4] But this meaning does not seem to make much sense in the biblical context of Numbers.[5]

The LXX, in Rahlfs' edition, gives as a translation: Ἀλλ' ἢ ἐν φάσματι δείξει Κύριος, 'But if the Lord will show in a wonder,' or something similar, since the meaning of φάσμα, as given in Liddell & Scott, varies from 'apparition, phantom, a sign from heaven, portent', to 'monster or prodigy.'[6] Other Greek manuscripts read ἐν χάσματι δείξει, which seems to me a simple (and later) adaptation to the context, what we could call a rendering *ad sensum*, χάσμα meaning precisely 'chasms, gulf, gapping mouth, or generally any wide opening,' and being thus quite well adapted to the story that follows. Maybe for this reason the *New English Bible* translates simply: 'But if the Lord makes a great chasm.'

Kittel's edition of the Biblia Hebraica proposes to correct the Hebrew according to the Greek and to read: בראי יראה 'will show in my seeing,' a conjecture retained with a question mark in the Stuttgartensia, although this edition is less prone to textual corrections than its predecessor, which indicates that the masoretic reading is indeed problematic. Unfortunately, this unique mention of בריאה is lacking in the Qumran manuscripts of Numbers, since not one manuscript has preserved the passage, which deprives us of the possibility of deciding on

---

[4] As does B.A. Levine, *Numbers 1–20* (AB 4A), Garden City, NY 1993, 408, who translates 'But if YHWH creates [a special] creation.'

[5] H.E. Hanson, 'Num. XVI 30 and the Meaning of bara',' *Vetus Testamentum* 22 (1972) 353–359, which proposes to give the verb ברא here the primary meaning of 'to cut,' and translates 'But if the Lord splits open a crevice.' See further J. Körner, 'Die Bedeutung der Wurzel bara im Alten Testament,' *Orientalistische Literaturzeitung* 64 (1969) 533–540, and M. Miguéns, 'BR' and Creation in the Old Testament,' *Liber Annuus* 24 (1974) 38–69; Miguéns traces the semantic development of the verb through the whole Hebrew Bible. R.J. Clifford and J.J. Collins, *Creation in the Biblical Traditions* (CBQ MS 24), Washington 1992, 140–142 contains a well-chosen bibliography on the topic.

[6] H.G. Liddell and R. Scott, *A Greek-English Lexicon: With a Revised Supplement*, Oxford 1996, 1919. *Le Pentateuque: La Bible d'Alexandrie*, Paris 2001, 382 translates: 'Seulement le Seigneur fera montre d'un prodige' and adds in a note: 'TM: "Si YHWH crée une création", peut être au sens d'une chose nouvelle, un prodige' (p. 775).

the matter.[7] Be that as it may, it is clear that this single use of the abstract substantive for 'creation' in the biblical texts is not without its problems. And we may conclude that the abstraction level we are looking for was not yet reached in the Biblical text.[8]

2. *Ben Sira and the targumim*

This level may be present in Ben Sira. In 16:16, MS A from the Geniza reads in Beentjes' edition:[9] **רחמיו יראו לכל בריותיו**, which is translated by Skehan & Di Lella as 'His mercy was seen by all his creatures.'[10] We find here **בריותיו**, the plural of **בריאה**, used (as in later rabbinic literature) to designate the results of the creative act of God, the creatures. But some uncertainty also remains in this case since this verse is absent from the Greek I and from the Latin translations, and we cannot confirm its antiquity nor exclude the intrusion of later vocabulary.[11]

The targumim, of course, have no problems with the meaning of the word in Num 16:30, and they understand it in the meaning the word **בריאה** has in rabbinical literature. They are, of course, too late to be of interest for our purpose, but a quick look at them, nevertheless, is helpful in understanding an important element of the interpretation of

---

[7] A total of 11 manuscripts of the Hebrew texts have been recovered from the different collections: 1Q**3** frg. 8–23 (1QpaleoNum) published by D. Barthélemy, in: *DJD* 1, 53–54; 2Q**6–9** (2QNum[a-d]) published by M. Baillet, in: *DJD* 3, 57–60; 4Q**23** (4QLev-Num[a]) published by E. Ulrich, in: *DJD* 12,153–176; 4Q**27** (4QNum[b]) edited by J. Nastram, in: *DJD* 12, 205–267; Mur **1** published by J.T. Milik, in: *DJD* 2, 78; 5/6Hev **1a,** XHev/Se **1** published by P. Flint, in: *DJD* 38, 137–140 and 173–177; and 34Se **2** published by M. Morgenstern, in: *DJD* 38, 209, as well as a copy of LXX Numbers, 4Q**121**, edited by P. Skehan, E. Ulrich, and J.E. Sanderson, in: *DJD* 9, 187–194.

[8] I do not think that the correction of MT **לברית** ('upon the covenant') of Ps 74:20 into **לבריות** ('upon the creation') proposed by Kittel is needed, nor Dahood's emendation into *l^e^birateka* ('upon your temple') [M. Dahood, *Psalms II: 51–100* (AB 17), Garden City, NY 1968, 208] since the Hebrew text makes perfect sense as it is.

[9] P.C. Beentjes, *The Book of Ben Sira in Hebrew: A Text Edition of all Extant Hebrew Manuscripts and A Synopsis of all Parallel Hebrew Ben Sira Texts* (VT Sup 68), Leiden 1997, 46.

[10] P.W. Skehan and A.A. Di Lella, *The Wisdom of Ben Sira* (AB 37), Garden City, NY 1987, 268.

[11] 'These two verses (15–16) are present in MS A, in Greek II, and in Syr; they are not vouched for by GI or Lat, and represent a late expansion of the text,' conclude Skehan & Di Lella, *The Wisdom of Ben Sira*, 270.

creation we will find in the scrolls: that many other things not expressly mentioned in the narrative of Genesis were also created in the first week.

Onqelos, as usual, remains the closest to the MT. Neofiti[12] translates 'But if the Lord creates *a new* creature,' adding thus the word חדתה, and reflecting the interpretation which was current in rabbinical circles and is already reflected in Jerome's translation in the Vulgate 'Sin autem *novam rem* fecerit dominus.' But in the margin of the Neofiti manuscript appears a completely different interpretation of the biblical text which can be rendered: 'If from the days of the world,[13] death was created in the world, behold it is a good thing for this world; if not, let it be created now; and let (the earth) open, ...'[14] The meaning of this interpretation is not obvious, but it seems to imply that the new thing created by God to which the main text alludes is nothing other than death, which if it had not been created by God from the beginning would have been created specially for the punishment of the two rebels. This is at least what the translation of Pseudo-Jonathan explicitly says,[15] of which Neofiti Margin in my view represents a garbled reflection:

> And if the death has not been created for them (לא איתבריית מיתותא להון), from the days of the (beginning of) the world it is created for them now (תתברי להון כדון), and if a mouth has not been created for the earth (פום לארעא) from its beginning, it is created for it now; and the earth shall open the mouth and shall swallow them up and all that is theirs, etc.[16]

Pseudo-Jonathan is here alluding to the well-known rabbinic midrash of the ten things that were created at twilight between the sixth day and the Sabbath,[17] a midrash he curiously does not place when translating

---

[12] A. Díez Macho, *MS. Neophyti 1. Tomo IV Números* (Textos y Estudios 10), Madrid 1974, 157.

[13] Or 'from eternity' (מן ומי [leg. יומי] עלמא).

[14] In the translation by M. McNamara, in: Díez Macho, *MS. Neophyti 1*, 562.

[15] I use the edition by T. Martínez Sáiz and A. Díez Macho (eds), *Biblia Polyglotta Matritensia. Series IV; Targum Palestinense in Pentateuchum. Additur Targum Pseudojonathan ejusque hispanica versio. L. 4 Numeri*, Madrid 1977, 157.

[16] In the translation of E.G. Clarke, in: M. McNamara and E.G. Clarke, *Targum Neofiti 1: Numbers. Targum Pseudo-Jonathan: Numbers* (The Aramaic Bible 4), Edinburgh 1995, 235.

[17] The enumeration of the ten things that were created at twilight occurs often in rabbinic literature and is found in at least five different forms depending on the way these marvellous things are grouped. In *Pirke Abot* 5:6 and *Pirke de Rabbi Eliezer* 19:1 'the mouth of the earth' is the first thing enumerated in the list. A.J. Saldarini, *The Fathers*

Num 16:32 (although he faithfully translates the expression partially omitted by the LXX, ופתחת ארעא ית פומה), but in Num 22:28 (the episode of the ass of Balaam, the last of the wonderful things created in its list):

> Ten things were created after the world was established, with the coming in of the Sabbath between the suns: the manna, the well, Moses' staff, the diamond, the rainbow, the clouds of Glory, the earth's mouth, the writing of the tables of the covenant, the demons, and the mouth of the speaking ass.[18]

3. *The Dead Sea Scrolls*

Thus, except for the dubious usage in Ben Sira, the earliest attestations of the use of the word that will become the standard word in later Hebrew to designate both the creative act by God (the singular noun בריאה, the creation) and the results of this creative act (its plural בריאות, the creatures) are found in the Dead Sea Scrolls.[19] Of course, the use of the verb ברא to indicate God's creative action is overwhelming within the Hebrew Bible, and the derivation of the abstract substantive from it poses no particular problems. We cannot thus rule out that the word was already used before the time of the Scrolls. But it is a curious fact that in the Hebrew Bible the abstract word 'creator' is never expressed with the participle of ברא, but with the participles of other roots, יצר (Jer 10:16; 51:19; Sir 51:12), עשה (Isa 22:11; 27:11; 44:2; 51:13; 54:5; Hos 8:14; Amos 4:13; 5:8; Ps 115:5; 121:2; 124:8; 134:3; 136:4; Job 4:17; 35:10; Prov 14:31; 17:5; 22:1), or פעל (Job 36:3) for example.

This may or may not explain why the word בריאה and its plural בריאות are not used. But the fact remains that in the literature known to us they are attested for the first time in the Scrolls. And this proves in my opinion that the level of abstraction we are looking for was already reached at Qumran.

According to Rabbi Nathan (Abbot de Rabbi Nathan Version B) (SJLA 11), Leiden 1975, 306–310 contains a very useful appendix on the different lists and show clearly that the lists of *ARNB* 37 and the list of Ps-Jon Num 22:28 'have no close similarity in groupings or overall pattern to any of the list or to each other.'

[18] Translation by Clarke, in: McNamara and Clarke, *Targum Neofiti 1: Numbers*, 254.

[19] M.G. Abegg, *The Dead Sea Scrolls Concordance: The Non-Biblical Texts from Qumran*, Leiden 2003, vol. I, 157–158 lists 52 occurrences of the verb ברא, and 23 occurrences of the noun בריא or בריאה, of which 18 in the singular and 6 in the plural (בריאות).

Therefore I will first offer an overview of the uses of (1) בריאות and (2) בריאה in the Scrolls. Afterwards I will examine (3) a short hymn included in 11QPsalms[a], most probably of non-sectarian origin, entitled *Hymn of the Creator*.[20]

3.1. הבריאות: *'the creatures'*

Let us start with the uses of הבריאות in the Scrolls.[21] Some of the uses cannot be exploited because of the lack of concrete context. In 4Q**181** 2:10,[22] for example, the last preserved word of the fragment, which because of the large margin could have preserved the beginning of the composition called *Ages of Creation*, is certainly בריאות, but the word is without context and incomplete. The editor reads it as if it had a feminine suffix בריאותיה 'her creatures,' but as Strugnell observes, the word could equally be read with a plural suffix בריאותיה]ם, 'their creatures.'[23] But we cannot say anything about its concrete meaning, nor ascertain to whom the suffix refers. Even less can be concluded from the single occurrence registered in the indexes of *4QInstruction* and translated by 'His creatures' by the editors.[24] The reading is most uncertain, and the form of the orthography of the word (without *yod*) equally uncertain.[25] In the rest of the occurrences, the meaning seems

[20] I will leave for another occasion the analysis of the use of the creation story of Genesis in a typical Qumran composition, the first column of the Sukenik edition of the *Hodayot* (column 9 in the new numbering of the *Hodayot*) and the study of three aspects of creation in the *Tractate of the Two Spirits*: the creation of man (1QS 3:17–18), the creation of the spirits (1QS 3:25) and the new creation (1QS 4:25).

[21] 4Q**181** 2:10; 4Q**216** 5:9; 4Q**266** 10 ii 10; 4Q**287** 3:2; 4Q**416** 1:17; 4Q**504** 1–2 recto vii 9.

[22] Edited by J. Allegro, *Qumran Cave 4.1* (DJD of Jordan 5), Oxford 1968, 79–80, plate XVIII.

[23] J. Strugnell, 'Notes en marge du volume V des "Discoveries in the Judaean Desert of Jordan",' *Revue de Qumran* 26/7 (1970) 163–276, esp. 255. This is the reading adopted by J.J.M. Robert, in: J.H. Charlesworth (ed.), *The Dead Sea Scrolls: Hebrew, Aramaic, and Greek Texts with English Translations*, vol. 2 (PTSDSSP 2), Tübingen & Louisville 1995, 208, who translates 'their creations.'

[24] 4Q**416** 1:17, edited by J. Strugnell and D.J. Harrington, *Qumran Cave 4. XXIV: Sapiential Texts, Part 2* (Discoveries in the Judaean Desert 34), Oxford 1999, 81, plate III. The editors register the word as such in the Concordance to the volume (p. 550), suggesting that the word is possibly 'a participle נבראתיו *Nip'al*; the *Nip'al* of ברא occurs once in the Bible, but at least three more times in the 1–11Q texts' (p. 88).

[25] The substantive is read by E.J.C. Tigchelaar, *To Increase Learning for the Understanding Ones: Reading and Reconstructing the Fragmentary Early Jewish Sapiential Text 4QInstruction* (STDJ 44), Leiden 2001, 176.

to be all inclusive, alluding to everything God has created, and we can consequently translate the plural as 'creatures.'

(a) Fragment 3 of one of the copies of 4QBerakoth (4Q**287** 3) has preserved part of an interesting blessing, which I quote in the translation of the editor, Bilha Nitzan:[26]

1. [in] their [awes]ome deeds, and they will bless Your holy name with blessings of [... the holiest of the holy ones]
2. [And] all creatures of flesh (בריאות הבשר), all those [You] created (כולמה אשר בראתה) [will ble]ss you ...
3. [... ca]ttle and birds and creeping things and fish of the [s]eas and all ... [
4. [... Y]ou created them all anew [... (אתה בראתה את כולמה מחדש).

This blessing, obviously based on the narratives of Genesis, and combining 1:24–26 with other passages as shown by the vocabulary, has several interesting features. It designates the creatures as such, using the plural בריאות, and specifically connecting this name with the creative act of God אשר בראתה. It designates all the creatures as 'creatures of flesh,' which could be read in the light of the peculiar meaning of 'flesh' in the Scrolls,[27] although in this case the expression could be simply a way of distinguishing these creatures from the angelic ones mentioned in the first line ('the holiest of the holy ones'), according to the editor, who thinks the fragment may overlap in this way with frg. 12 of 4Q**286.** The fragment specifies further that these creatures have been created anew (מחדש), which could also be interpreted in the light of the new creation theology (the עשות חדשה of 1QS 4:25, for example), but which most probably means simply that creation is the very beginning of the creatures, or may allude to the creatures born after the flood, as Nitzan suggests.

(b) In column seven of the first copy of the composition known as *4QWords of the Luminaries* (4Q**504** 1–2 VII),[28] we find a prayer specifically

[26] In E. Eshel *et al.* (eds), *Qumran Cave 4. VI: Poetical and Liturgical Texts, Part 1* (DJD 11), Oxford 1998, 54.

[27] See J. Frey, 'The Notion of Flesh in 4QInstruction and the Background of Pauline Usage,' in: D.K. Falk, F. García Martínez, and E. Schuller (eds), *Sapiential, Liturgical and Poetical Texts from Qumran* (STDJ 35), Leiden 2000, 197–226.

[28] Edited by M. Baillet, *Qumrân Grotte 4. III* (DJD 7), Oxford 1982, 150. E. Puech offered a reconstruction of the manuscript on the basis of the shape of the fragments

composed to be said on the Sabbath, and designated in the manuscript both as 'Hymns' (הודות) and as 'song' (שיר). Only the right part of the column has been preserved. In the translation of the *Study Edition* (vol. 2, 1017) we can read:

> Praise, /A song/ for the Sabbath day. Give thanks …
> his holy Name for ever …
> all the angels of the holy vault (מלאכי רקיע קודש) and …
> to the heavens, the earth and all its schemers (לשמים הארץ וכול מחשביה)
> …
> the great abyss, Abaddon, the water and all that there [is in it …
> all its creatures (כול בריאותיו) always, for centuries [eternal (תמיד לעולמי עד).
> Amen. Amen.

Here, too, the reference to the Genesis narrative is evident. Here, too, the angels are present, associated expressly with the רקיע. Here, too, are the 'creatures' as a collective designation, although the preceding lacuna has deprived us of knowing to whom the third person singular pronoun refers. It is particularly unfortunate that the reading מחשביה is uncertain. We have translated it (with Baillet) as from חשב, and understood it as referring to the earthly schemers[29] because the word appears in 1QH 11:32–33, a passage which has several word links with ours, although in this description of the eschatological destruction the word has a clear negative connotation, which is not apparent here.[30] But the word could equally well be read מחשכיה and interpreted as the dark places of the earth, from חשך, as in Ps 74:20: 'for the dark places (מחשכי) of the land are full of the haunts of lawlessness.' In this text the references to the creatures are also linked to the action of giving thanks to God.

---

and the formulaic content in his review of *DJD* 7 in *Revue Biblique* 95 (1988) 404–411. See also E. Chazon, 'On the Special Character of Sabbath Prayer: New Data from Qumran,' *Journal of Jewish Music and Liturgy* 15 (1992/93) 1–21, and D.K. Falk, *Daily, Sabbath, and Festival Prayers in the Dead Sea Scrolls* (STDJ 27), Leiden 1998, 59–94.

29 'Être pensants' in the translation by Baillet, in: *DJD* 7, 151.

30 In the translation of the *DSSSE*, vol. 1, 167: 'It consumes right to the great abyss. The torrents of Belial break into Abaddon. The schemers of the abyss (מחשבי תהום) howl at the din of those extracting mud. The earth cries out at the calamity which overtakes the world, and its schemers (מחשביה) scream, and all who are upon it go crazy, and melt away in the great calamity.'

(c) The plural בריאות also appears in one of the copies of the *Damascus Document*, where its editor translates it with 'people.' But in my opinion this restrictive meaning is unwarranted. In 4Q**266** 10 ii 9–10 we can read in the translation of the editor:[31]

> He who goes about naked in the house in the presence of his fellow, or out in the field in the presence of people (הבריאות),[32] shall be excluded for six months.

The first part of this text is a clear parallel of one of the specifications of the penal code of the *Serek* (1QS 7:12): 'And whoever walks about naked in front of his fellow, without needing to, shall be punished for six months.' The *Damascus Document* distinguishes two sorts of transgressions, one done in the house and the other outside the house, in the field; the one is done in front of the fellow (רעהו) and the other in front of the הבריאות. Baumgarten seems to have been guided in his restrictive interpretation of the word by the way the sentence appears in another copy of the *Damascus Document*, 4Q**270** 7 i 2,[33] where the text reads בבית או בשדה, apparently abbreviating the full expression of 4Q**266** and mixing up one of the two transgressions.[34] But this interpretation forgets that in the same manuscript, the sentence has been corrected by an addition above the line of which enough has been preserved to assure us that the abbreviation of the sentence was nothing more than a copyist's mistake.[35] The restriction of the meaning of הבריאות to 'human beings' instead of the more general 'creatures' seems thus unjustified. What the texts intend to punish is exhibitionism (going around naked) in all circumstances. The restrictive clause of the *Serek* ('without needing to') has disappeared in the *Damascus Document*, and now it is punishable to go naked both inside and outside, not only in front of fellow members but also in front of any 'creature.'

(d) הבריאות also appears in one of the copies of the *Book of Jubilees* (4QJub[a]) found in Cave 4: 4Q**216** v 9.[36] The Ethiopic text of *Jub.* 2:2

[31] J.M. Baumgarten, *Qumran Cave 4. XIII: The Damascus Document* (DJD 18), Oxford 1996, 74.

[32] The reading is uncertain since the *bet* is missing (see *DJD* 18, plate XIII), but the reconstruction seems assured.

[33] *DJD* 18, 162.

[34] *DJD* 18, 75.

[35] See *DJD* 18, plate XXXIV.

[36] Edited by J.C. VanderKam and J.T. Milik, in: H. Attridge *et al.* (eds), *Qumran Cave 4. VIII: Parabiblical Texts Part 1* (DJD 13), Oxford 1994, 13.

contains a long list of the angels which were created on the first day of the creation, and concludes with a summary statement indicating that besides all the specified angels, 'the spirits of all his creatures which are in heaven and in earth' were also then created.[37] Two fragments of 4Q**216**, frgs. 12 ii and 13[38] allow the editors to reconstruct most of *Jub.* 2:2, including the summary. The key words 'the spirits of his creatures' (רוחות הבריותו) have been preserved and can be read clearly in line 4 of frg. 12 ii, but the space to be filled to the next preserved word in frg. 13 is a little too large for the Ethiopic text. Therefore the editors reconstruct: '[all] the spirits of his creatures [what he made in the heavens, which he made on the ear]th, and in every (place).' As in the previously quoted texts, 'creatures' here apparently refers to all the results of the creative work of God. In the Ethiopic version, the phrase could be understood as referring both to humans and to celestial phenomena like the winds, the clouds, the thunder, which as humans have a 'spirit' (רוח). But the specification of וכול in the Hebrew original (although grammatically a little awkward)[39] indicates clearly that the plural הבריות 'creatures' was already used as the common designation for all things God (to whom the pronoun refers) had created. To the best of my knowledge, the plural is only used with this meaning in the texts of Qumran. The singular הבריאה, is never used to designate a single 'creature,' but as we shall see is used to designate the creative act of God, the 'creation.'

3.2. הבריאה: *'the creation'*

The word הבריאה appears fourteen times in the manuscripts, sometimes only partially preserved and always with the determinative.[40] We can group these occurrences into three categories: when it is used as a temporal reference, when it is used a-temporally to express the results of God's activity, and when it is applied to a future reality.

---

[37] See O.H. Steck, 'Die Aufname von Genesis 1 in Jubiläen 2 und 4. Esra 6,' *Journal for the Study of Judaism* 8 (1977) 154–182; J.C. VanderKam, 'Genesis 1 in Jubilees 2,' *Dead Sea Discoveries* 1 (1994) 300–321; J.T.G.A.M. van Ruiten, *Primaeval History Interpreted: The Rewriting of Genesis 1–11 in the Book of Jubilees* (JSJ Sup 66), Leiden 2000, 20–27.

[38] *DJD* 13, plate I.

[39] In the *DSSSE* we have translated it by 'and in everything.'

[40] CD 4:21 (and partially in the parallel passage from 6Q**15** 1,3); 4Q**216** 5:1; 4Q**217** 2:2; 4Q**223–224** 43:4; 4Q**225** 1:7 bis; 4Q**226** 1:8; 4Q**253** 2:3; 4Q**267** 1:8; 4Q**319** 4:11; 4:17; 4Q**320** 3 i 10; 11Q**19** 29:9.

a. *'Creation' as a reference to the very beginnings of things*
הבריאה is sometimes used as a *temporal reference*, a shorthand expression to indicate the very beginnings of things. This is the meaning which the word has in one of the Commentaries on Genesis, *4QCommentary on Genesis B* (4Q**253**).[41] In frg. 2:3 we read טהורים מן הבריאה, which, if we take into consideration the next line which speaks of 'holocaust for acceptance,' should be translated as 'pure (animals) from the creation' as we have done in the *Study Edition*,[42] rather than as 'pure things from creation' as the editor does. Independent of the plural subject of which the purity is asserted, these texts clearly affirm that it has this quality from the very beginning, 'from the creation.'

The word appears with the same meaning (as a temporal reference) in two closely related texts: 4QOtot (4Q**319**)[43] and 4Q**320**, one of the *Mishmarot* or Calendars,[44] and the same meaning is the most logical when it appears in a manuscript closely related to the *Book of Jubilees* from Cave 4 (4Q**217**),[45] which may be a similar and closely related composition, or another copy of the same *Book of Jubilees* in which the fragment preserved does not correspond exactly to the Ethiopic text.[46]

Notwithstanding its separate publication as an independent manuscript, 4Q**319** is part of one of the copies of the *Rule of the Community* found in Cave 4 (4QS^{e}), where its contents replaced the section known as the 'Hymn of the Maskil' in the copy from Cave 1.[47] The beginning text of 4QOtot is fragmentary, but nevertheless the first three lines (col. IV, lines 9–11 in the *DJD* edition)[48] can be read:

---

41 Edited by G. Brooke, in: G. Brooke *et al.*, *Qumran Cave 4. XVII: Parabiblical Texts, Part 3* (DJD 22), Oxford 1996, 209–212.

42 *DSSSE*, vol. 1, 505.

43 Edited by J. Ben-Dov, in: Sh. Talmon, J. Ben-Dov, and U. Glessmer, *Qumran Cave 4. XVI: Calendrical Texts* (DJD 21), Oxford 2001, 319–245.

44 Edited by Sh. Talmon and J. Ben-Dov, in: Talmon, Ben-Dov, and Glessmer, *Qumran Cave 4. XVI*, 35–63.

45 Edited by J.C. VanderKam and J.T. Milik, in: H. Attridge *et al.* (eds), *Qumran Cave 4. VIII: Parabiblical Texts Part 1* (DJD 13), Oxford 1994, 23–33.

46 For these texts, see the contribution by E.J.C. Tigchelaar in this volume.

47 See S. Metso, *The Textual Development of the Qumran Community Rule* (STDJ 21), Leiden 1997, 48–51. The editors of 4QS^{e} also conclude: 'The fragments of 4QS^{e} and of the calendrical work known as *4QOtot* (4Q**319**) belong to the same scroll. 4QS^{e} ends in col. IV with text corresponding to 1QS IX 20–24 and *4QOtot* follows immediately' (*DJD* 26, 131).

48 *DJD* 21, 214.

[...]blessed [...]
[...] its light on the fourth day of the wee[k ...]
[... the] creation (הבריאה) in the fourth (day) in Ga[mul ...].

The reading of בריאה in 4:11 is practically certain, even if the *bet* is partially covered by a fold of the leather.[49] In the same column, in 4:17, the editors read again the word בריאה, but its occurrence here seems to me less assured on palaeographical grounds. In the calendrical text 4Q**320** we find another mention of הבריאה, this time undisputed, in frg. 3 i 10. This fragment, which preserves the left part of a column, reads, according to the editors:[50]

[...] the years of holiness
[... the] holy creation (הבריאה קדש)
[... on the fo]urth day in the week
[of Gam]ul, head of all the years (רוש כול השנים)
[ot]ot of the second jubilee.

These two texts obviously echo Gen 1:14–19, the basic reference text for all calendrical texts from Qumran, and more concretely 1:14 when, on the fourth day of creation God creates the luminaries to divide the day from the night, and placed them לאתת ולמועדים ולימים ושנים 'for signs and for festivals and for years and for days.' On the fourth day of creation, which is here defined as 'holy,' the luminaries start pouring out their light, allowing the measurement of time, the days, years and jubilees, and this absolute beginning ('the head of all the years') can be traced exactly back using the *mishmarot* system. It happened during the week of Gamul. Because of the fragmentary character of these texts, it remains uncertain whether the luminary they are speaking about is the sun or the moon. But in my opinion, there is no doubt that the moon is the protagonist, as I proposed in an article years ago based on the beginning of 4Q**320**, frg. 1 i 1–5.[51] Precisely for this passage, the editors of *DJD* 21 reconstruct another mention of הבריאה. The text reads in their translation:[52]

[...] to its being seen (or: appearance) from the east
[...] to shine in the middle of the heavens at the foundation of

[49] The *bet* is clear in PAM photograph 41.479.
[50] *DJD* 21, 50.
[51] F. García Martínez, 'Calendarios en Qumrán (II),' *Estudios Bíblicos* 54 (1996) 523–552.
[52] *DJD* 21, 42–45.

> [creatio]n from evening until morning on the 4th (day) of the week (of service)
> of Gamul in the first month of the first (solar)
> year.

That this text is speaking of the moon is certain, since the luminary in question shines 'from evening until morning.' The editors reconstruct the expression ביסוד [הבריא]ה, basing themselves on the parallel found in the *Damascus Document* that we will discuss below, and asserting that the two strokes (which are the only preserved part of the word) should be read as *he* because they are parallel and perpendicular according to PAM photograph 41.700.[53] But in the oldest photograph of the fragment (PAM 40.611), the first stroke joins the second at a clear angle, making the reading of the remains as an *ayin* the most logical solution, which makes the reading of the *Study Edition* ביסוד [הרקי]ע 'at the base of the [vaul]t' the more plausible.[54] With or without a new mention of הבריאה, this text (with its clear reference to the fourth day of the creation in Gen 1:14–18) confirms our understanding of the use of the word in the other two calendrical texts as a clear temporal reference to the beginning of the days.

This temporal reference appears in its clearest form in the passage from 4Q**217** 2:2.[55] This text has been published as a possible copy of the *Book of Jubilees* (4QpapJubilees[b]?), but the few elements that have been preserved do not have exact correspondence in the Ethiopic text.[56] The temporal meaning of 'the creation' as a temporal reference point, sandwiched as it is between the two temporal prepositions מן ... עד, is so clear that it needs no comment. The editors' translation reads:[57]

> [...] the divisions of the times for the law and for the [testimony ...]
> [...]for all the ye[ars of] eternity, from the creatio[n ...] (מן הבריאה)
> [...]*m* and all [that has been] created until the day wh[ich ...] (עד היום).

---

[53] *DJD* 21, 43. In the photograph which they publish as Plate I (PAM 43.330), only one of the two strokes is visible. The other is very faint, but certainly not perpendicular.

[54] *DSSSE*, vol. 2, 679.

[55] See note 45. Equally clear is the temporal reference of the expression יום הבריאה, but its precise meaning is rather complex and needs to be dealt with separately below.

[56] *DJD* 13, 24: 'The contents of the text are largely unknown, but it does mention the years (frg. 1,2), מחלקות העתים which is the Hebrew name of *Jubilees* (frg. 2,1), the day of creation, and possibly Jerusalem (2,4). The contents of the second fragment resemble what is said in *Jub.* 1:26–29.'

[57] *DJD* 13, 25–26.

b. *'Creation' as the result of God's activity*

The word הבריאה in other texts does not have any temporal connotations and it simply expresses the results of God's activity. One example of this use is found in one of the copies of *Jubilees* from Cave 4, 4QJub[a], already quoted in the first section.[58] In 4Q**216** 5:1, which closely corresponds to the beginning of chap. 2 of the Ethiopic book, Moses is ordered to write 'all the words of the creation' (כול דברי הבריאה). Which is exactly what Moses does, of course, and in a very detailed way when compared with the masoretic text.

The two instances in which the word הבריאה appears in the *Damascus Document*[59] also lack any temporal connotation in my view.

In CD 4:21 we read: ביסוד הבריאה זכר ונקבה ברא אותם.[60] What is important in our present perspective is not the quote from Gen 1:27 'male and female he created them,' but the explicit assertion that this is the יסוד הבריאה, an expression that can be translated as 'the principle of creation,'[61] or 'the foundation of creation.'[62] From the perspective of the text and its defence of monogamy, we could even translate the whole sentence as 'the essence of creation is: they were created as a single male and single female.' The exegetical point is clearly the singular form of both ונקבה and זכר, and since the phrase is taken from Gen 1:27, where Eve has not yet been mentioned, the singularity of man and woman is considered essential to the human race.[63] Creation is not seen here as a temporal marker of the beginning of mankind,[64] but as an

---

[58] See above, notes 36 and 37.

[59] Edited by S. Schechter, *Documents of Jewish Sectaries. 1. Fragments of a Zadokite Work*, Cambridge 1910. A better transcription with excellent photographs is the one prepared by E. Qimron, in: M. Broshi (ed.), *The Damascus Document Reconsidered*, Jerusalem 1992, 9–49. J.M. Baumgarten and D.R. Schwartz have produced a new edition in: J.H. Charlesworth (ed.), *The Dead Sea Scrolls: Hebrew, Aramaic, and Greek Texts with English Translations*, vol. 2: *Damascus Document, War Scroll and Related Documents* (PTSDSSP 2), Tübingen & Louisville 1995, 4–57.

[60] The phrase has not been preserved in any of the copies of the *Damascus Document* found in Cave 4, edited by J.M. Baumgarten, *Qumran Cave 4. XIII: The Damascus Document (4Q266–273)* (DJD 18), Oxford 1996, but is partially present in the copy from Cave 6 edited by Milik, in: M. Baillet, J.T. Milik, and R. de Vaux, *Les 'Petites Grottes' de Qumrân* (DJD of Jordan 3), Oxford 1962, 129.

[61] As we have done in *DSSSE*, vol. 1, 557.

[62] Which is the translation given by Baumgarten & Schwartz, *Damascus Document*, 19.

[63] See F. García Martínez, 'Man and Woman: Halakhah based upon Eden in the Dead Sea Scrolls,' in: G.P. Luttikhuizen (ed.), *Paradise Interpreted* (TBN 2), Leiden 1999, 95–115.

[64] In *DJD* 21, 45 the editors explicitly assert the temporal aspect of ביסוד הבריאה they reconstruct in 4Q**320**, in my opinion without any basis.

expression of its nature:[65] God has created mankind sexed, and from this characteristic follows that a man cannot take two wives.

The other occurrence of the word in CD 12:14–15[66] does not refers to humans but to the locusts or grasshoppers of Lev 11:22 (החגבים), and does not use יסוד but משפט. The text reads כי הוא משפט בריאתם, where the plural suffix refers back to the החגבים במיניהם, and the meaning of the sentence is similar to the expression used in CD 4:21. In the translation in the *Study Edition*:[67]

> And all the locusts, according to their kind, shall be put into fire or into water while they are still alive, as this is the regulation for their species.[68]

Locusts come in many sorts. Lev 11:22 mentions specifically four kinds: הארבה, הסלעם, החרגל and החגב. We cannot be precise about to which sort of locust each name corresponds, but it is clear that in later times החגב has become a sort of collective name for all these species. In rabbinic literature, החגב is the word used almost exclusively to designate locusts. In *m. Terumot* 10:9 and *m. Eduyot* 7:2, for example, pure and impure החגבים are distinguished, while in *m. Abodah Zara* 2:7 its suitability for consumption depends on where one has bought them—those from the stock of a shop are suitable whilst those from the [shopkeeper's] basket are not. *m. Hullin* 3:7 shows clearly that in rabbinic times החגב has become the collective name for locust.[69] Apparently, this was already the case at Qumran, since, except when clearly quoting or alluding to biblical texts,[70] החגב is the only name used for locust.

---

65 C. Rabin, *The Zadokite Fragments*, Oxford 1958, 16–17 translates 'the principle of nature' and adds in a note: 'Perhaps this is nothing but a translation of φύσις.'

66 Partially preserved in *4QDamascus Document*[a], 4Q**266** 9 ii 1–2, see *DJD* 18, 68.

67 *DSSSE*, vol. 1, 571.

68 Literally: 'because this is the norm of their creation.' Rabin, *The Zadokite Fragments*, 62, translates 'for this is what their nature requires,' and Baumgarten & Schwartz, *Damascus Document*, 53: 'for this is the precept of their creation.' C. Hempel, *The Laws of the Damascus Document: Sources, Traditions and Redaction* (STDJ 29), Leiden 1998, 160 offers a translation *ad sensum*: 'for this is how they are to be eaten.'

69 In *m. Hullin* 3:7 we can read: 'And among locusts (ובחגבים): Any which have (1) four legs, (2) four wings, and (3) jointed legs [Lev 11:21], and (4) the wings of which cover the greater part of its body. R. Yose says, "And (5) the name of which is locust (ושמו חגב)"' (transl. by J. Neusner, *The Mishnah: A New Translation*, New Haven 1988, 772).

70 As is the case in *11QTemple* (11Q**19** 48:3–4) and *4QReworked Pentateuch* (4Q**365** 15a–b, 5) which quote Lev 11:22 and thus use the different names for locusts, and of *4QParaphrase of Genesis and Exodus* (4Q**422** 3:5) which retells the story of the plagues of Egypt, using both ארבה from Exod 10:4 and חסיל from Ps 78:46 (written חסל in the manuscript); see E. Tov, 'The Exodus section of 4Q422,' *Dead Sea Discoveries* 1 (1994) 197–209.

Also in CD 12:15, the temporal connotation of הבריאה is absent; the text underlines that the nature of each species is fixed from their creation. God has created different sorts of locusts, and this determines the way they should be consumed, boiled in water or roasted on the fire.[71] But not cooked in milk, a practice permitted by the rabbis according to *m. Hullin* 8:1.[72]

c. *'Creation' as reference to a future reality*

At Qumran we also find two occasions where although the word הבריאה has a clear temporal dimension, it does not seem to refer to the past 'creation' but is applied to a future reality, a 'creation' which has not yet taken place.

In the *Temple Scroll*[73] 29:8–9, Yadin read the expression עד יום הברכה in the following phrase:

> I shall sanctify my temple with my glory, for I shall make my glory reside over it *until the day of blessing*, when I shall create my temple establishing it for myself for ever, in accordance with the covenant I made with Jacob at Bethel.

The phrase has certain difficulties, of course, but it makes perfect sense. The temple for which the *Temple Scroll* legislates is not the final one. On the 'day of blessing' God himself will create a new one which will be the definitive temple and will endure forever. But Qimron[74] proposed reading the key expression as עד יום הבריה, where הבריה is a different way of writing הבריאה. From a palaeographical point of view, this reading is, if not certain, at least clearly to be preferred. And if this reading is accepted, the 'day of creation' in question cannot refer back to the first creation, since this 'day of creation' is clearly in the future, and still out of sight. This is made clear by the future form of the verb used (אברה) and especially by the use of עד 'until.' Are we dealing here

---

71 Rabin, *The Zadokite Documents*, 62, explains that the locust should be put into fire or water while they are alive because 'they were created from these elements,' and notes that drawing the locusts in water is demanded by the Samaritans and Karaites.

72 כל הבשר אסור לבשל החלב חוץ מבשר דגים וחגבים 'Every [kind of] flesh [of cattle, wild beast, and fowl] is prohibited to be cooked in milk, except for the flesh of fish and locust.' Ibidem, 780.

73 Y. Yadin, *Megillat ham-Miqdash—The Temple Scroll*, 3 vols + suppl., Jerusalem 1977 (Hebrew), 1983 (revised English edition).

74 First in E. Qimron, 'The Language of the Temple Scroll,' *Leshonenu* 42 (1978) 136–145 (Hebrew), on page 142, and later in *The Temple Scroll: A Critical Edition with Extensive Reconstructions* (Judean Desert Studies), Beer Sheva & Jerusalem 1996, 44.

with an *Urzeit-Endzeit* typology? Or has יום הבריאה become one of the designations of eschatological time?

That both elements may have been combined in this expression seems proved by the only other occasion when the expression is used, in one of the manuscripts called *Pseudo-Jubilees*: 4Q**225** 1:6–7.[75] There we can read:[76]

> [...] *vacat* And you, Moses, when I speak with [you ...]
> [...] the creation until the day of the [new] creation [...] (הבריאה עד יום הבריאה [החדשה]).

As so often happens in Qumran research, the frustrating fragmentary state of the manuscript precludes all certainty. But in this case, the editor VanderKam has build a very strong case to complete the sentence in the light of *Jub.* 1:27 and 1:29.

In *Jub.* 1:27 God tells the angel of the presence to dictate to Moses the events '... from the beginning of creation until the time when my temple is built among them throughout all the ages of eternity.' And in *Jub.* 1:29, which is textually garbled but is usually restored according to the suggestion of M. Stone, we read:[77] 'from [the time of creation until] the time of the new creation when the heavens, the earth, and all their creatures shall be renewed ...' Based on these two quotations, VanderKam most plausibly suggests completing the broken sentence of our manuscript with the word החדשה. The temporal connotation of the 'day of creation' is retained, and the 'day of the new creation' also takes on a temporal connotation (עד), which is clearly situated in the eschatological future, the moment of the renewal of everything: 'the heavens, the earth and all their creatures.' As the *Temple Scroll* shows, this moment, this 'new creation,' could simply be called יום הבריה 'the day of creation.'

At Qumran the word 'creation' is not only used to express the creative act of God, or the temporal beginning of the reality created by God, but also the expected renewal of the reality ('the heavens, the earth, and all their creatures') in the eschatological future. The 'day of creation' is not only the model of the end times but one of its names as well.

---

[75] Edited by J.C. VanderKam and J.T. Milik, in: H. Attridge *et al.* (eds), *Qumran Cave 4. VIII: Parabiblical Texts Part 1* (DJD 13), Oxford 1994, 141–155, plate X. See also R. Kugler and J.C. VanderKam, 'A Note on 4Q225,' *Revue de Qumran* 20/77 (2001) 109–115.

[76] *DJD* 13, 143–144.

[77] M. Stone, 'Apocryphal Notes and Readings,' *Israel Oriental Studies* 1 (1971) 125.

3.3. *Hymn to the Creator*

Since my specific topic was the interpretation of the biblical narrative of the creation at Qumran, I cannot close without giving at least one example of how some of its elements were developed in the writings found there. I have selected a short poem,[78] not particularly original,[79] which has been transmitted in one of the manuscripts (11Q5 or *11QPsalms*[a]) of what many authors consider to be a 'Qumran Psalter,' different from the masoretic Psalter in the ordering of the Psalms included, and with several other known or previously unknown compositions included as part of it.[80] Column 26 of this manuscript, after the end of Psalm 149 and Psalm 150, has preserved this little poem almost completely[81] in lines 9–15, which I quote in the versified translation of its editor J. Sanders:[82]

---

[78] Edited by J.A. Sanders, *The Psalms Scroll of Qumran Cave 11* (DJD of Jordan 4), Oxford 1965, 47. The Hymn has attracted the attention of scholars and the number of studies dedicated to it is rather great: J. Carmignac, 'Le texte de Jérémie 10,13 (ou 51,16) et celui de 2 Samuel 23,7 améliorés par Qumrân,' *Revue de Qumran* 26/7 (1970) 287–290; P. Skehan, 'A Liturgical Complex in 11QPs[a],' *Catholic Biblical Quarterly* 35 (1973) 195–205; P. Skehan, 'Jubilees and the Qumran Psalter,' *Catholic Biblical Quarterly* 37 (1973) 343–347; F. García Martínez, 'Salmos Apócrifos en Qumrán,' *Estudios Bíblicos* 40 (1982) 197–220; M. Weinfeld, 'The Angelic Song over the Luminaries in the Qumran Texts,' in: D. Dimant and L.H. Schiffman (eds), *Time to Prepare the Way in the Wilderness* (STDJ 16), Leiden 1995, 131–157; K. Seybold, 'Das Hymnusfragment 11QPs[a] XXVI 9–15,' in: Idem, *Studien zur Psalmenauslegung*, Stuttgart 1998, 188–207; G. Xeravits, 'Notes sur le *11QPs*[a] Creat 7–9,' *Revue de Qumran* 69/18 (1997) 145–148; E.G. Chazon, 'The Use of the Bible as a Key to Meaning in Psalms from Qumran,' in: S. Paul *et al.* (eds), *Emanuel: Studies in Hebrew Bible, Septuagint, and Dead Sea Scrolls in Honor of Emanuel Tov* (VT Sup 94), Leiden 2003, 85–95.

[79] In the words of the editor (*DJD* 4, 89): 'The metre is highly irregular, and the language is forced and pedestrian. The imagery and vocabulary are late, in biblical terms.' Carmignac, 'Le texte de Jérémie,' 287 defined it as 'en fait un centon de citations bibliques.'

[80] See P. Flint, *The Dead Sea Psalms Scrolls and the Book of Psalms* (STDJ 17), Leiden 1997 and U. Dahmen, *Psalmen- und Psalter-Rezeption im Frühjudentum: Rekonstruktion, Textbestand, Struktur und Pragmatik der Psalmenrolle 11QPs*[a] (STDJ 49), Leiden 2003.

[81] See Seybold, 'Das Hymnusfragment,' 199. In the reconstruction by Dahmen, only one and a half lines separate the Hymn from the following text from 2 Sam 23:1–7; see Dahmen, *Psalmen- und Psalter-Rezeption*, 96 and 249. Carmignac, 'The Texte of Jérémie,' 289 is the only scholar who considers that the Hymn ended in the next column with the text of 2 Sam 23:7, quoted by the author of the poem in a similar way to the way he had already quoted the text of Jeremiah.

[82] *DJD* 4, 47, 89–91.

Great and holy is the Lord,
the holiest unto every generation.
Majesty precedes him,
and following him is the rush of many waters.
Grace and truth surround his presence;
truth and justice and righteousness are
the foundation of his throne.

Separating light from deep darkness,
by the knowledge of his mind he established the dawn.
When all his angels had witnessed it they sang aloud,
for he showed them what they had not known:
Crowing the hills with fruit,
good for every living being.

Blessed be he who makes the earth by his power,
establishing the world in his wisdom.[83]
By his understanding he stretched out the heavens,
and brought forth [wind] from his st[orehouses].
He made [lightning for the rai]n,
and caused mist[s] to rise [from] the end of the [earth.]

The structure of the poem is quite simple, with three sections, each composed of three stanzas: the description of God's glory, its own description of His act of creation, and the transformation of the traditional description of creation into the closing blessing.

The most characteristic element of the poem is its reuse of materials attested elsewhere. And this phenomenon is evident in the three components of the hymn. The most obvious and best studied is the presence of the 'floating' piece we know from its double appearance in the book of Jeremiah, in 10:12–13 and 51:15–16, and from its presence in Ps 135:6–7, which is reused with little variation in the last section of our poem.[84]

E. Chazon, in her very illuminating article,[85] has shown how the author of the hymn has transformed Jeremiah's text with the simple introduction of the word ברוך, the omission of the difficult לקול תתו (also omitted by the LXX in Jer 10:13, but not in Jer 28:16, which corresponds to the Hebrew 51:16) and the transposition of the line 'and

---

[83] In the light of the exegetical traditions which apparently underlie the poem, I would rather translate '*with* his wisdom,' as we have done in the *DSSSE*, vol. 2, 1179.

[84] Xeravits, 'Notes,' proposes an order of dependence Ps 135 > Jeremiah > Qumran, as against Sanders who postulated a Ps 135 > Qumran > Jeremiah relationship. Chazon, 'The Use of the Bible,' 92 assumes that our text quotes Jeremiah directly.

[85] Chazon, 'The Use of the Bible,' 90–94.

brought forth [wind] from his st[orehouses].' In this way the author of the poem deeply transforms a prophecy of doom (against Israel in Jeremiah 10 and against Babylon in Jeremiah 51) into blessing and thanksgiving.

Chazon also suggests reading the first section in the light of the *merkavah* visions of Ezekiel 1 and Isaiah 6. She even entertains the idea that the three repetitions of the word *holy* in the first bicolon[86] could be an echo of the *trishagion* and that the hymn is witness to the 'praying with the angels' tradition, though she does not agree with the suggestion by Weinfeld that the Hymn would preserve an ancient form of the *Qedushah* liturgy.[87] But in the light of the targumic texts quoted at the beginning (of the many things that were created before creation), and of the *Jubilees* text about the creation of the angels on the first day, also quoted above, I am also more inclined to see the opening stanza as a witness to the very old exegetical traditions which try to solve the problems posed by the irregularities in the text of the biblical narrative.[88]

This is clearer in the second section which elaborates on the biblical narrative of the creation (already used in Jeremiah and in the Psalm), and where several of the basic components of the narrative of Genesis are cleverly incorporated: the separation of light and darkness, the heavens and the earth, and all its produce as nourishment for every living being.

The presence of the angels need not refer to the Ezekiel or Isaiah descriptions. For our poet, as for the author of *Jubilees*, angels were already present since the first day of creation. *Jubilees* reaches this conclusion from the presence of the רוח אלהים in Gen 1:2.[89] Our text

---

[86] Of which a literal translation would be: 'Great and *holy* is YHWH, the *holy* of the *holiest* [or the holiest of the holy ones] for generation to generation.'

[87] Weinfeld, 'The Angelic Song,' has suggested that our hymn reflects an ancient form of the morning liturgy, the *Qedusah Yoser*.

[88] See P. Schäfer, 'Berešit Bara' 'Elohim: Zur Interpretation von Genesis 1,1 in der rabbinischen Literatur,' *JJS* 2 (1971) 161–166; F. García Martínez, 'Interpretación de la creación en le Judaísmo antiguo,' and M.L. Sánchez León (ed.), *La Creació* (Religions del món antic 2), Palma 2001,115–135. For a synthetic treatment of the exegetical developments, see G. Vermes, 'Genesis 1–3 in Post-Biblical Hebrew and Aramaic Literature before the Mishnah,' *JJS* 43 (1992) 221–225; M.D. Goulder, 'Exegesis of Genesis 1–3 in the New Testament,' *JJS* 43 (1992) 226–229; P. Alexander, 'Pre-Emptive Exegesis: Genesis Rabba's Reading of the Story of Creation,' *JJS* 43 (1992) 230–245.

[89] See Van Ruiten, *Primaeval History Interpreted*, 25. Rabbinic tradition, which places the creation of the angels on different days of the creation week usually deducts its creation (of which the biblical text is completely silent) from the צבאם of Gen 2:1, which is read as the creation of the angelic 'hosts.'

does not make explicit the exegetical 'peg' but firmly asserts that they were created before the world was and were present and acclaiming God's creation.

Nor does the presence of God's heavenly throne need to lead us to Ezekiel or Isaiah. God's heavenly throne is one of the seven things created before the world's creation, according to *b. Pesah* 54a, for example. The exegetical conclusion could have been derived from Ps 93:1–2 'your throne stands firm from of old,' and it is attested in *2 Enoch* 25:3–4. The personification of the divine attributes can also be exegetically explained, and even more easily the creation of the world through God's wisdom, its 'establishment' as our text calls it.

That God established the world with Wisdom, was also exegetically acquired from old. At least since the קנני ראשית דרכו, which starts the description of Wisdom in Prov 8:22 and implies considering ראשית to be one of the names of Wisdom (as Philo explicitly says), and leads to the translation of the בראשית of the biblical text, as 'together with Wisdom' created God, etc.[90] We find this translation in the Fragment Targum, and as a double translation in Neofiti.[91] The presence of Wisdom, and co-operation in the creation work, will also be gratefully used to explain the plurals of the creation of man, of course, and in the Christian tradition it will lead to the involvement of the Son in the creation, both via his identification with Wisdom (as in the *Letter of Barnabas* 5:5), or through his identification with the Logos of John 1:1 as in Col 1:13–15 or Heb 1:2.[92]

The article by Skehan[93] brought to light a parallel between the beginning of the second section of our poem and the book of *Jubilees* 2:2–3.[94] Both texts share the phrase 'Separating light from darkness he established the dawn by the knowledge of his heart' in very similar or identical wording.[95] They also seem intended to solve a well-known exegetical problem: the creation of light on the first day and

---

[90] B.L. Mack, *Logos und Sophia: Untersuchungen zur Weisheitstheologie im hellenistischen Judentum*, Göttingen 1973.

[91] See G. Anderson, 'The Interpretation of Gen.1:1 in the Targums,' *Catholic Biblical Quarterly* 52 (1990) 21–29.

[92] J.L. Moreno Martínez, 'El Logos y la creación,' *Studia Theologica* 15 (1983) 381–419.

[93] P. Skehan, '*Jubilees*, and the Qumran Psalter,' *Catholic Biblical Quarterly* 37 (1975) 343–347.

[94] Partially preserved in 4Q**216** 5:10–11, translated by the editors: 'darkness, dawn, [light, and evening which he prepared through] his knowledge. Thus we saw his works and we [blessed him], regarding all his [wo]rks' (*DJD* 13, 14).

[95] Van Ruiten, *Primaeval History Interpreted*, 26 offers a useful synopsis of the two

the posterior creation of the luminaries only on the fourth day (see also Tigchelaar, this volume). The light of the luminaries, although not yet created on the first day, was already established in God's mind.

Jubilees is not the only composition which has been influenced by this poem. *4QAdmonition Based on the Flood* (4Q**370**)[96] has reworked the last part of the second section of our hymn. The phrase 'Crowning the hills with fruit, good food for every living being,' has been transformed into 'And he crowned the mountains with produce and poured our food upon them. And with good fruit he satisfied all'.[97] And it is perhaps also significant that 4Q**370** follows the phrase quoted with an exhortation to blessing: 'Let all who do my will eat and be satisfied' said YHWH. 'And let them bless my holy name.' Although here worded as divine speech, and possibly motivated by the presence of the three verbs 'eat, be satisfied, and bless' in the same order as Deut 8:10, this exhortation to blessing corresponds with the introduction of the same verb that is used by the poet in order to transform the meaning of the old 'floating' piece of Jeremiah and Psalm 135.

I will certainly not claim that such a later interpretation as the one in *y. Hag.* 2:77c, in *b. Ber.* 55a or in *Genesis Rabbah* 1:10 on the creation of the world by using the letter *bet*, or the one so beautifully worded in the medieval midrash known as *Alphabet of Rabbi Akiva* of the contest of the letters for getting the honour to be the first of the Torah,[98] is already present in our text. But the addition of ברוך, which transforms the quotation into a blessing, as well as the correspondences between creation and blessing noted when dealing with the word הבריאה, seems to indicate that at the time of the composition of this poem, the בריאה was already linked with the ברכאה.

texts, underlining the similarities and the differences, and concludes that our hymn has influenced the author of *Jubilees*.

[96] Edited by C. Newsom, in: M. Broshi *et al.* (eds), *Qumran Cave 4. XIV: Parabiblical Texts Part 2* (DJD 19), Oxford 1995, 85–97.

[97] Newsom shows clearly the dependence: 'Although the priority of 11QPs[a] cannot be independently demonstrated, it appears that the author of 4Q**370** has cited the first colon of 11QPs[a] XXVI 13 in a slightly adapted form, and then paraphrased or expanded each of the following terms of the second colon in order to create his own text. Thus אוכל becomes ו]שפך אכל על פניהם; טוב becomes ופרי טוב השביע; and לכול חי becomes כל נפש or 'כל נפש כל אשר עשה רצוני' (*DJD* 19, 91–92).

[98] See L. Ginzberg, *The Legends of the Jews*, Philadelphia 1968[12], vol. 1, 5–8, where the *bet* wins the contest, using as argument that all humanity shall bless God continuously through it.

PART II

CREATION IN ANCIENT GREECE, ANCIENT PHILOSOPHY, AND THE EARLIEST GRAECO-ROMAN INTERPRETATIONS OF GENESIS 1

CANONICAL AND ALTERNATIVE CREATION MYTHS IN ANCIENT GREECE

Jan N. Bremmer

Introduction

Near Eastern peoples normally heard about the creation on ritual occasions:[1] *Illuyankaš* was the cult legend of the Hittite Purulli festival;[2] *Enuma elish* was recited during the Babylonian New Year festival (*ANET* 331–334), and Egyptian cosmogonic myths were alluded to every day in the hymns sung during the temple rituals, which themselves were re-enactments of the creation.[3] In Greece, on the other hand, poems with (fragmentary) accounts of the creation could be performed at festivals but also at the courts of kings and aristocrats. In my contribution I will first present a brief analysis of what I call the canonical versions of the creation, since they occur in Homer and Hesiod, the traditional teachers of ancient Greek religion (§ 1). Secondly, I will discuss two accounts influenced by Orphism, a somewhat later, alternative current within Greek culture (§ 2). I will conclude with a discussion of the possibly Persian origin of the epigrammatic formulation of Genesis 1:1 (Appendix).

1. *Canonical versions*

1.1. *Homer*

Whereas Near Eastern cosmogonic myths reach back at least into the second millennium, ancient Greece came rather late to its cosmogonies, just as Israel must originally have lacked a full cosmogony, given its virtual absence from Ugarit (below). Local Greek histories show that

---

[1] For good surveys of creation accounts see H. Schwabl, 'Weltschöpfung', in: *Real-Encyclopädie der classischen Altertumswissenschaft*, Suppl. IX (1962) 1433–1589; A. Merkt *et al.*, 'Weltschöpfung', in: *Der Neue Pauly* XII.2 (2002) 463–474.

[2] See the beginning of the myth of Illuyanka in: H. Hoffner, Jr, *Hittite Myths*, Atlanta 1990, 11; J.V. García Trabazo, *Textos religiosos hititas*, Madrid 2002, 82–83.

[3] See S. Sauneron and J. Yoyotte, 'La naissance du monde selon l'Égypte ancienne',

traditionally the beginning of the world was presupposed, although anthropogonies did occasionally exist.[4] It was a sign of the rise of Greek civilization and its growing contacts with the Near East that in the eighth century BC poets started to borrow from the Near East to fill this gap. The first attempts are still visible in the *Iliad.* In a passage that has recently been repeatedly discussed, Hera announces that she wants to reconcile 'Okeanos, begetter of the gods, and mother Tethys' (XIV.201). The English Prime Minister William Gladstone, who was highly interested in the contemporary discoveries of cuneiform tablets,[5] already realized that this couple derived from the beginning of *Enuma elish*, where we read:

> When skies above were not yet named
> Nor earth below pronounced by name,
> Apsu, the first one, their begetter
> And maker Tiamat, who bore them all,
> Had mixed their waters together,
> But had not formed pastures, nor discovered reed-beds;
> When yet no gods were manifest,
> Nor names pronounced, nor destinies decreed,
> Then gods were born within them' (I.1–9).[6]

Walter Burkert, to whose studies my contribution is much indebted, has convincingly shown that the Greek Tethys is a perfect transcription of Akkadian Tiamat.[7] This means that Okeanos is the Greek version of Apsu. His etymology is clearly un-Greek,[8] and his origin is obscure,

---

in: *La Naissance du monde* (Sources orientales 1), Paris 1959, 17–91; S. Bickel, *La cosmogonie égyptienne avant le Nouvel Empire*, Fribourg and Göttingen 1994.

[4] J.N. Bremmer, 'Pandora or the Creation of a Greek Eve', in: G.P. Luttikhuizen (ed.), *The Creation of Man and Woman*, Leiden 2000, 19–33.

[5] W.E. Gladstone, *Landmarks of Homeric Studies*, London 1890, 129–132. For his presence during the announcement by George Smith of the discovery of the Flood on a Gilgamesh tablet, see Bremmer, 'Near Eastern and Native Traditions in Apollodorus' Account of the Flood', in: F. García Martínez and G. Luttikhuizen (eds), *Interpretations of the Flood*, Leiden 1998, 39–55 at 39.

[6] All translations of Mesopotamian myths are from S. Dalley, *Myths from Mesopotamia*, Oxford 2000².

[7] W. Burkert, *The Orientalizing Revolution*, Cambridge, Mass. 1992, 92–93 and *Die Griechen und der Orient*, Munich 2003, 37 (to be read with the reservations of M.L. West, *The East Face of Helicon*, Oxford 1997, 147 note 20), accepted by R. Janko on *Iliad* XIV.200–207.

[8] For his pre-Greek name see E.J. Furnée, *Die wichtigsten konsonantischen Erscheinungen des Vorgriechischen*, The Hague 1972, 124; W. Fauth, 'Prähellenische Flutnamen: Og(es)-Ogen(os)-Ogygos', *Beiträge zur Namenforschung* 23 (1988) 361–379; West, *East Face*, 146–148.

but the fact that the strange epithet *apsorhoos* is applied only to him (XVIII.399) suggests that Homer realised the resemblance. Okeanos is the fresh water that encircles the world and the source of all rivers and springs (XXI.195–197). The couple appears several times in Greek mythology, as in Hesiod and in an Orphic poem quoted by Plato in his *Cratylus*: 'The handsome river Okeanos was the first to marry, he who wedded his sister Tethys, the daughter of his mother'.[9] Here the couple keeps its primacy,[10] which was even appropriated by local mythology. In a fragment of his *Corinthiaca* the poet Eumelus mentions that Corinth took its alternative name from 'Ephyra, the daughter of Okeanos and Tethys' (frg. 1b Davies = 1 Bernabé).[11] On the other hand, according to Plato in the *Timaeus* (40E), Okeanos and Tethys are the children of Ouranos and Gaia, but parents of Kronos and Rhea. Apparently, the couple itself was canonical, but their place in the divine genealogy could vary somewhat.

Does this mean that the Homeric mention of the couple was also part of an old theogony, as Richard Janko has suggested?[12] That seems doubtful. As our quotation from *Enuma elish* shows, the couple has been taken from the beginning of that poem. However, this passage was not the only Greek borrowing from Near Eastern literature. The casting of the lots by Zeus, Poseidon and Hades in the *Iliad* (XV.187–193) was also derived from the beginning of a Near Eastern poem, the Akkadian *Atrahasis*. Similarly, the Hittite *Song of Kumarbi*, from which Hesiod, directly or indirectly, borrowed the castration of Ouranos, is the first song of the *Kumarbi Cycle*.[13] Evidently, the early Greeks took some of their Oriental material from the beginning of the great Near Eastern epics, poems that were especially popular in school curricula.[14]

---

[9] Hesiod, *Theogony* 337, 362, 368, frg. 343.4 MW; Acusilaus, frg. 1 Fowler; Plato, *Cratylus* 402B = *Orphicorum Fragmenta* [henceforth: *OF*], frg. 15 Kern [henceforth: K] = 22 Bernabé [henceforth: B].

[10] For the problems of this verse see most recently, if not totally persuasively, M.L. West, *The Orphic Poems*, Oxford 1983, 120; A. Bernabé, *Hieros logos: Poesía órfica sobre los dioses, el alma y el más allá*, Madrid 2003, 56.

[11] For the *Corinthiaca* see now M.L. West, 'Eumelus', *Journal of Hellenic Studies* 122 (2002) 109–133 at 118–126.

[12] Janko on *Iliad* XIV.200–207.

[13] H. Hoffner, 'The Song of Silver', in: E. Neu and C. Rüster (eds), *Documentum Asiae Minoris Antiquae*, Wiesbaden 1988, 143–166.

[14] Burkert, *Orientalizing Revolution*, 95; for another example add now J.H. Huehnergard and W.H. van Soldt, 'A Cuneiform Lexical Text from Ashkelon with a Canaanite Column', *Israel Exploration Journal* 49 (1999) 184–192. We should perhaps also note that P. Michalowski, 'The Libraries of Babel: Text, Authority, and Tradition in Ancient

We need therefore not postulate an elaborate pre-Homeric theogony or cosmogony. A simple allusion is all that will have been there.

Now what did this first cosmogony mean to the early Greeks? Just like the Babylonians with Tiamat and Apsu and the Egyptians with their couple Nun and Naunet, the primordial waters,[15] some Greeks apparently imagined the beginning of the world as water; the idea prefigures Pherecydes' (frg. 64 Schibli) and Thales' (A 12 Diels-Kranz) idea of water as the first principle (see also Adams, this volume, §2).[16] The Israelites clearly did the same: 'the spirit of God hovered upon the face of the waters' (Gen 1:2). The mention of *tôhû* in this verse points to Tiamat and shows that both the Israelites and the Greeks borrowed the idea of water as primeval element from the Mesopotamians. In Greece Okeanos thus replaced and expanded the function of the river god Acheloôs, the former Greek origin of all the world's waters.[17]

There is also another Mesopotamian element in the Iliadic cosmogony. Okeanos is called 'begetter of the gods'. This, too, must have come from the same first verses of *Enuma elish*, but such a detail was naturally rejected by the Israelites. It remains noteworthy that any reference to the beginning is lacking in this ultra-short Greek cosmogony. There is no *bereshit* (Gen 1:1), the word that refers to *Enuma elish*'s calling Apsu 'the first one', *reshtu*, a word of the same root. *Archê* will appear only with Thales and the like.

But was this primordial couple really the very first? In the passage from the *Iliad* in which it occurs we also meet Night, who apparently occupies such an important position that even Zeus dares not offend her (XIV.261).[18] Later poets and philosophers repeatedly mention Night as the first principle. Night already concludes Hesiod's 'reversed cos-

---

Mesopotamia', in: G. Dorleijn and H. Vanstiphout (eds), *Cultural Repertories*, Leuven 2003, 105–129 at 118 observes that many Near-Eastern libraries had only one or two tablets of the great epics.

[15] S. Morenz, *Ägyptische Religion*, Stuttgart 1960, 184.

[16] Thales (= Aristotle, *Metaphysica* 983b20), cf. U. Hölscher, *Anfängliches Fragen*, Göttingen 1968, 9–89; J. Rudhardt, *Le thème de l'eau primordiale dans la mythologie grecque*, Berne 1971; O. Keel, 'Altägyptische und biblische Weltbilder, die Anfänge der vorsokratischen Philosophie und das *Archê*-Problem in späten biblischen Schriften', in: B. Janowski and B. Ego (eds), *Das biblische Weltbild und seine altorientalischen Kontexte*, Tübingen 2001, 27–63.

[17] See the splendid demonstration by G. d'Alessio, 'Textual Fluctuations and Cosmic Streams: Ocean and Acheloios', *Journal of Hellenic Studies* 124 (2004). I am most grateful to Giambattista d'Alessio for showing me a copy of his article before publication.

[18] C. Ramnoux, *La Nuit et les enfants de la Nuit dans la tradition grecque*, Paris 1959, 62–108 ('La nuit de la cosmogonie') is not really helpful.

mogony' in his *Theogony* (11–20). She is also the first element in Musaeus (B 14 Diels-Kranz), together with Aer in Epimenides (B 5 Diels-Kranz = frg. 6 Fowler), and in early Orphism, as we now know from the Derveni papyrus (XIV.6 = *OF* 10B) where Night is the mother of Heaven. Night was also first with Silence in Antiphanes' comedy *Theogony*.[19] She figured in Anaxagoras (B 51 Diels-Kranz) and Chrysippus (*apud* Philodemus, *De pietate* 359–360 Obbink),[20] and is mentioned as such by Aristotle (*Metaphysica* 1071b26–27) and his pupil Eudemos (frg. 150 Wehrli²);[21] after two introductory hymns, the imperial collection of *Orphic Hymns* also starts with a hymn to Night.[22]

Darkness is of course another frequent characterization of the primeval situation before the actual creation, although we do not find primeval darkness in Mesopotamia. The Roman mythographer Hyginus, who summarized and compiled Greek traditions, started his *Fabulae* with a strange hodgepodge of Greek and Roman cosmogonies and early genealogies. It begins as follows: *Ex Caligine Chaos. Ex Chao et Caligine Nox Dies Erebus Aether* (*Praefatio* 1). His genealogy looks like a derivation from Hesiod, but it starts with the un-Hesiodic and un-Roman 'Darkness'. Darkness probably did occur in a cosmogonic poem of Alcman,[23] but it seems only fair to say that it was not prominent in Greek cosmogonies.

The situation must have been different in Egypt. Here we encounter the primeval couple Kek and Keket, 'Darkness' in the *Ogdoad* of Hermopolis,[24] and in *Genesis* we read that 'darkness was upon the face of the deep' (1:2). In this connection it is interesting to note that Philo of Byblos also calls the primeval situation 'dark air and slimy chaos' (*FGrH* 790 frg. 2). Admittedly, he ascribed his information to Sanchuniathon 'from

[19] Cf. R. Kassel and C. Austin, *Poetae comici Graeci*, vol. 2, Berlin & New York 1991, 366–367.

[20] Cicero, *De natura deorum* 3.44, who mentions a Stoic genealogy that began with Erebos and Night, probably refers to this text.

[21] *OF* 24, 28, 28a, 310K = 20B, cf. L. Brisson, 'Damascius et l'Orphisme', in: Ph. Borgeaud (ed.), *Orphisme et Orphée en l'honneur de Jean Rudhardt*, Geneva 1991, 157–209 at 201–202, repr. in Brisson, *Orphée et l'Orphisme dans l'Antiquité gréco-romaine*, Aldershot 1995, VI.201–202. For Eudemos' fragment see now G. Betegh, 'On Eudemus Frg. 150 Wehrli', in: I. Bodnár and W.W. Fortenbaugh (eds), *Eudemus of Rhodes*, New Brunswick & London 2002, 337–357.

[22] L. Robert, *Opera minora selecta*, vol. 7, Amsterdam 1990, 569–573, with a discussion of, surely Orphic, dedications to Night.

[23] Alcman, frg. 81.21 Calame = 5.frg. 2 iii.21 Davies, cf. S. Rangos, 'Alcman's Cosmogony Revisited', *Classica et Mediaevalia* 54 (2003) 81–112 at 93–94.

[24] Morenz, *Ägyptische Religion*, 184.

before the Trojan War', but we may well wonder whether Philo did not borrow from the Jews in this respect:[25] he certainly also borrowed from Orphic cosmogony.[26] Moreover, there was no proper Ugaritic cosmogony,[27] and Philo's primeval mother Baau, interpreted by him as Night, has long been compared with *Genesis*' *tôhû wa-bôhû* (1:2).[28] Darkness is indeed still the primeval situation in the *Strassbourg cosmogony*, the Late Antique Greek poem from, almost certainly, also Hermopolis. It is related to Hermetic literature and describes how Hermes (here the *interpretatio Graeca* of Thoth) creates the world.[29] It is not impossible, then, that Homer derived Night too, directly or indirectly, from the beginning of a Near Eastern cosmogonical account.[30]

What I would like to conclude from this discussion so far is that archaic Greece already displayed an interest in cosmogony, but that some poets seem to have gone no further than making brief observations, which they, directly or indirectly, derived from the Ancient Near East. These observations still presuppose a biomorphic mode, to

[25] A.I. Baumgarten, *The* Phoenician History *of Philo of Byblos*, Leiden 1981, 144 objects to the interpretation of Baau's partner Kolpia, 'Wind', as deriving from *ruah*, 'wind' in Gen 1:2, 'as it makes Philo's sources too Biblical to be true'. This is evidently no valid argument, as Philo actually wrote a *Peri Ioudaiôn* (*FGrH* 790 frgs. 9–11)! For the demonstrable influence of *Genesis* on Philo see M.J. Edwards, 'Philo or Sanchuniathon? A Phoenician Cosmogony', *Classical Quarterly* 41 (1991) 213–220 at 217–218; K. Koch, 'Wind und Zeit als Konstituenten des Kosmos in phönikischer Mythologie und spätalttestamentlichen Texten', in: M. Dietrich and O. Loretz (eds), *Mesopotamica-Ugaritica-Biblica*, Neukirchen-Vluyn 1993, 59–91.

[26] This was already noted by William Robertson Smith (1846–1894), in the second and third series of his famous lectures on *The Religion of the Semites* (1890–1891), which have only recently been published, cf. W.R. Smith, *Lectures on the Religion of the Semites (Second and Third Series)*, ed. J. Day, Sheffield 1995, 104–107. For Philo's borrowing from the Greeks see also Edwards, 'Philo or Sanchuniathon?'; F. Millar, *The Roman Near East 31 BC – AD 337*, Cambridge, MA 1993, 277–278; G. Bowersock, *Fiction as History*, Berkeley, Los Angeles, London 1994, 43–44.

[27] A.S. Kapelrud, 'Creation in the Ras Shamra Texts', *Studia Theologica* 34 (1980) 1–11; G. Casadio, 'Adversaria Orphica et Orientalia', *Studi e Materiali di Storia delle Religioni* 52 (1986) 291–322; S. Ribichini, 'Traditions phéniciennes chez Philon de Byblos: une vie éternelle pour des dieux mortels', in: C. Kappler (ed.), *Apocalypses et voyages dans l'au-delà*, Paris 1987, 101–116.

[28] Baumgarten, *The* Phoenician History *of Philo*, 145; West, *Orphic Poems*, 188.

[29] See most recently G. Fowden, *The Egyptian Hermes*, Princeton 1993^2, 175; K. Geus, 'Ägyptisches und Griechisches in einer spätantiken Kosmogonie', in: K. Döring *et al.* (eds), *Antike Naturwissenschaft und ihre Rezeption*, vol. 8, Trier 1998, 101–118 (with thanks to Klaus Geus for kindly sending me a copy of his article).

[30] For this primeval position of Night note also that it is the oldest owner of the Delphic oracle, cf. the discussion by C. Sourvinou-Inwood, *'Reading' Greek Culture*, Oxford 1991, 242 note 73.

use Burkert's terminology, in which creation is following the model of genealogical myth, whereas in *Genesis* the technomorphic mode—'God created …'—is more prominent.

1.2. *Hesiod*

For a full genealogy, though, we have to turn to Hesiod. His *Theogony* also contains a cosmogony, but since that subject is not very dramatic, he presents it only in a nutshell. It may be useful to first give a translation:

First of all Chasm came into being; but next
wide-breasted Earth, always safe foundation of all
immortals who possess the peaks of snowy Olympus
and dark Tartara in a recess of the wide-pathed earth,
and Eros, who is fairest among immortal gods,
looser of limbs, who conquers the mind and sensible thought
in the breasts of all gods and all men.
From Chasm were born Erebos and black Night;
from Night were born Aither and Day,
whom she conceived and bore, joined in love with Erebos.
Earth first brought forth equal to herself
starry Heaven to cover her all over, and
to be an always safe home for the blessed gods.
She bore tall mountains, pleasing homes of divine
Nymphs who dwell in the valed mountains.
She also bore the barren sea, violent in surge,
Pontos, without love's union; but next
she lay with Heaven and bore deep-whirling Okeanos,
and Koios and Kreios and Hyperion and Iapetos,
and Theia and Rhea and Themis and Mnemosyne
and gold-crowned Phoibe and attractive Tethys.
After them was born the youngest, crafty Kronos,
most terrible of children; he hated his lusting father.[31]

Unlike *Enuma elish* and *Genesis*, Hesiod's world is gradually built from the bottom upwards. The parent of them all is Chaos, literally 'Chasm', in fact, a kind of Black Hole (814), even though it is not that easy to understand what Hesiod really meant by it. Epicurus even turned to philosophy because his schoolmaster could not explain to him its

[31] Hesiod, *Theogony*, 116–137, transl. R. Caldwell, *Hesiod's Theogony*, Cambridge, MA 1987, 35–37 (slightly adapted). In my discussion I am much indebted to the brilliant commentary by M.L. West, *Hesiod: Theogony*, Oxford 1966.

meaning in the *Theogony* (Diogenes Laertius 10.2). The gender of Chaos is neuter, as befits the absolute beginning, although it gives birth to Erebos and Night (123). From a passage later in the poem (740) we learn that it was situated between Earth and Tartarus, the deepest area of the underworld.[32] That is presumably why these two are mentioned next. The coming into being of Earth naturally reminds us of *Genesis*, where at the very beginning the earth is also already present, if 'without form and void' (1:2). Earth's primeval role is celebrated in her *Homeric Hymn* as 'mother of everything' and 'the oldest' (1–2). Yet she was not particularly honoured in ancient Greece. We know of only a few cults for her, and her ritual deviated from that of the Olympians. Apparently, her primeval role differentiated her from the later Olympians who had a much more developed personality.[33] The mention of Tartarus is less readily explicable, and some Greek authors, such as Plato and Aristotle, ignored lines 118–119.[34] Martin West (*ad loc.*) even thinks it possible that Tartarus was inserted as a Hesiodic afterthought. However, several early authors did mention Tartarus at the beginning of their cosmogony; Musaeus (B 14 Diels-Kranz) even started his creation story with him.[35] As Hesiod built his universe from the bottom up, to start with Tartarus seems fully understandable.

The next to be mentioned is Eros. His place here prefigures the quasi-demiurgic function that he occupies in early philosophers, poets and mythographers (see §2 below). In any case, it is a remarkable invention by Hesiod that finds no parallel in any of the other Ancient Near Eastern creation stories; the corresponding position of Pothos in Phoenician cosmology looks like one more derivation from the Greek tradition by Eudemos (frg. 150 Wehrli^2) or his source. Hesiod is followed only by the fifth-century Argive mythographer Acusilaus (*FGrH* 2 frg. 5 = frg. 6a,b Fowler) in not giving Eros any parents. Later authors, like Sappho (frg. 198 Voigt), Alcaeus (frg. 327 Voigt) and Simonides (*PMG* 575 Page), provided him with different parents, but their variations confirm the absence of an authoritative tradition in this respect.

It is only after these indispensable elements that the creation proper seems to take off. Chaos now gives birth to Erebos and Night, whereas

[32] See Bremmer, *The Rise and Fall of the Afterlife*, London & New York 2002, 4, 91.

[33] F. Graf, *Nordionische Kulte*, Rome 1985, 360; S. Georgoudi, 'Gaia/Gê. Entre mythe, culte et idéologie', in: S. des Bouvrie (ed.), *Myth and Symbol*, vol. I, Athens 2002, 113–134.

[34] See the discussion by West *ad loc.*; G. Kirk, J. Raven and M. Schofield, *The Presocratic Philosophers*, Cambridge 1983^2, 35.

[35] Epimenides, *FGrH* 457 frg. 4 = frg. 6a Fowler; Aristophanes, *Birds*, 693.

in *Genesis* God first creates light (1:3). Only subsequently are Aither and Day born from Night, but to make this process work properly, Hesiod had to change the grammatical gender of Erebos from neuter to masculine. In both cases, we first hear about the general categories, darkness and light, whereas only later do Night and Day arrive into the world.

It is perhaps surprising that Earth is mentioned only now as giving birth to Heaven, but God also created heaven rather late in *Genesis* (1:7–8). In Homer, heaven is called 'bronze' (XVII.425) and 'iron' (15.329), and it seems to have been represented as a solid roof, flat and parallel to the earth, as it is 'equal to her'. This symmetry is typical of Greek cosmologies: 'it is assumed that the great divisions of the world are of equal size and at equal distances apart' (West *ad loc.*). Heaven was an insignificant god, who had no cult in ancient Greece. That is perhaps why Hesiod stresses that heaven is the seat of the gods, who are normally located on Olympus.

It is a rather archaic element in this cosmogony that mountains are seen as something different from the rest of the earth. There may well be a trace of Hittite influence here, since a *Kumarbi* fragment states: 'seven times he sent me against the dark earth … and seven times he sent me against the heaven … and seven times he sent me against the mountains and rivers' (*KUB* XXXIII.105, transl. West). However, Marduk created mountains from Tiamat's udder (*Enuma elish* V.57), and mountains are also mentioned separately in God's creation in *Proverbs* (8:23–26) and considered to belong to the oldest elements of the creation (Ps 90:2, Job 15:7). The collocation of the rough mountains with the lovely Nymphs is a subtle touch in this cosmogony.[36]

After heaven, earth and mountains, we finally reach the sea, a good indication of the low esteem in which it was held by the Greeks.[37] Pontos is an obscure figure, not mentioned by Homer and without any cult;[38] similarly, in *Genesis* (1:10) the sea is mentioned virtually at

---

[36] For the creation of mountains in Chinese, Greek, Near Eastern and Islamic cosmology see now D. Accorinti, '*Parturiunt montes an parturiuntur?* La nascità delle montagne nel mito', in: idem and P. Chuvin (eds), *Des Géants à Dionysos: Mélanges offerts à F. Vian*, Alessandria 2003, 1–24.

[37] A. Lesky, *Thalatta*, Vienna 1947; E. Vermeule, *Aspects of Death in Early Greek Art and Poetry*, Berkeley, Los Angeles, London 1979, 179–209; R. Buxton, *Imaginary Greece*, Cambridge 1994, 99–101.

[38] For a possible representation see J. Doerig, 'Der Dreileibige', *Mitteilungen des Deutschen Archäologischen Instituts: Athenische Abteilung* 99 (1984) 89–95.

the end of God's creation of the universe. With the birth of the sea we come to the end of the 'immaculate conception' of Earth's children Heaven, mountains and the sea. It is not immediately clear why these have been produced without a father, but it seems that their 'primeval' status did not yet make them fit for being the product of a civilized marriage.

It is only now that Earth enters into a sexual relationship. With Heaven she brought forth the twelve Titans. West (*ad loc.*) comments that 'the marriage of Earth and Sky is a very common mythological motif'. It is certainly true that in Greece too the growth of nature was represented as a fertilization of earth by the rain of heaven. A good example is a fragment from Aeschylus' *Danaids*: 'Holy sky passionately longs to make love to earth, and desire (the Hesiodic Eros!) takes hold of earth to achieve this union. Rain from her bedfellow heaven falls and fertilizes earth, and she brings forth for mortals pasturage for flocks and Demeter's livelihood'.[39] Yet the Greek and Latin parallels of this fragment use the sexual relationship between Heaven and Earth only as a metaphor.[40] In fact, none of them proves that the Greeks saw nature as the product of a proper relationship between Heaven and Earth.

Moreover, the outcome of this sexual meeting goes in a completely different direction. When we look at the children produced, we see a rather mixed bunch. Admittedly, they are known collectively as the Titans, but only a few of them are really fitting for a cosmogony. Most have been taken by Hesiod from other contexts.[41] For our purpose we need to observe only that, unlike the passage from the *Iliad* with which we started, Okeanos and Tethys do not here form the first cosmogonic couple but are incorporated into the set of children. It cannot be true that, as West observes (*ad loc.*), the couple eventually suggest the separation of the upper and lower waters, a kind of parallel to the

[39] Aeschylus, frg. 44, 1–5 Radt, transl. Kirk, Raven and Schofield, *Presocratic Philosophers*, 39 (slightly adapted).

[40] Cf. Euripides, frgs. 839, 898, 941 Nauck^2; Lucretius 1.250, 2.992; Vergil, *Eclogues* 7.60 and *Georgics* 2.324ff., 991ff.; Horace, *Epodes* 13.2; Statius, *Silvae* 1.2.185–186; Plutarch, *Moralia* 770A; *Pervigilium Veneris* 59ff.

[41] For a detailed discussion see my 'Remember the Titans', in: C. Auffarth and L. Stuckenbruck (eds), *The Fall of the Angels*, Leiden 2004, 35–61; add now A. Bernabé, 'Autour du mythe orphique sur Dionysos et les Titans: Quelques notes critiques' and J.L. Lightfoot, 'Giants and Titans in Oracula Sibyllina 1–2', in: *Mélanges Vian* (note 35), 25–39 and 393–401, respectively.

separation of Heaven and Earth. There is no indication for such a meaning in the Greek or Mesopotamian texts.[42] On the other hand, we do notice that Hesiod had completely minimized the significance of the couple in his genealogy. There is a quiet polemic with Homer going on in his passage that can hardly be overheard.

2. *Alternative versions*

With the Titans we have come to the end of our discussion of the archaic cosmogonies. It was especially the so-called Orphic movement that was not satisfied with the solution that the poets had offered about the coming into being of universe and man. From about 500 BC onwards they offered alternative versions, although these did not carry the same authority as those by Homer and Hesiod. The most surprising find in this context is undoubtedly the Derveni papyrus, which has supplied us with the oldest original Orphic theogony.[43] This text has already received much attention in recent years, and that is why I would like to concentrate here on two other, shorter texts, one serious and one more humorous, that can give us some idea of this speculative movement and its concerns.

2.1. *Euripides*

In Euripides' tragedy *Wise Melanippe*, which probably dates from the 420s, the eponymous heroine says: 'Heaven and earth were once a single form; but when they were separated from each other into two, they bore and delivered into the light all things: trees, winged creatures, beasts reared by the briny sea—and the human race'. Her audience must have been pretty surprised to hear these doctrines, especially after she had assured them: 'This account is not my own; I had it from my mother'.[44] Kirk, Raven and Schofield take the latter information at face

[42] But notice that W. Burkert, *Kleine Schriften*, vol. II: *Orientalia*, Göttingen 2003, 235 compares Anaximander A 10 Diels-Kranz; Leucippus A 1 §32 Diels-Kranz.

[43] See the new, still preliminary, edition with translation by R. Janko, 'The Derveni Papyrus: An interim Text', *Zeitschrift für Papyrologie und Epigraphik* 141 (2002) 1–62.

[44] Euripides, frg. 484 Nauck^2 = J. Diggle, *Tragicorum Graecorum fragmenta selecta*, Oxford 1998, 122, transl. C. Collard, M.J. Cropp and K.H. Lee, *Euripides: Selected Fragmentary Plays*, vol. I, Warminster 1995, 253, whose commentary (mainly by Cropp) I have gratefully used.

value, as if Greek mothers would delight their children with cosmogonies, but this, surely, says more about English educational ideals than Greek practices.[45] The passage is quoted with other Orphic fragments on a Late Antique bowl (*OF* 66B), which assures its Orphic character.[46]

The theme of the separation of heaven and earth is also found in an Orphicising passage in Apollonius of Rhodes, where heaven, earth and sea together start off in one form.[47] The same idea occurs on a papyrus in which Zeus himself acts as demiurge (*SH* 938 Lloyd-Jones and Parsons = *OF* 68B) and in a fragment that was once ascribed to Democritus (B 5, 1 Diels-Kranz), but hardly seems to deserve it. Apparently, there was an Orphic tradition about the oneness of the primeval *materia*, even though we do not find this particular tradition in any extant Orphic poem.

But where did the idea derive from? Cornford noted already the antiquity of the motif and pointed to parallels in Indian, Babylonian, Egyptian and Chinese mythology, but went no further than this enumeration.[48] Cropp (*ad loc.*), on the other hand, specifies that the separation of heaven and earth is the ancient mythical conception found in the *Enuma elish*, although, we may add, this conception was derived from the Babylonians' western neighbours. Yet there it is said that Marduk used half of the slain Tiamat 'to roof the sky' (IV.135–146) and the other half to make the earth for gods and humans (V.61–62), which is not quite the same.[49] The tradition that heaven and earth were formed from an egg, as recorded in a Phoenician cosmogony recorded by the Greek Laitos, is hardly a convincing parallel either.[50] So, where does this tradition derive from?

---

[45] *Contra* Kirk, Raven and Schofield, *Presocratic Philosophers*, 43.

[46] For the Orphic content see A. Bernabé, 'Orphisme et Présocratiques: bilan et perspectives d'un dialogue', in: A. Laks and C. Louguet (eds), *Qu'est-ce que la Philosophie Présocratique?*, Lille 2002, 205–247 at 216–217.

[47] Apollonius Rhodius 1.494–511 = *OF* 29K = 67B.

[48] F. Cornford, *From Religion to Philosophy*, London 1912, 67.

[49] Th. Jacobsen, 'The Battle between Marduk and Tiamat', *Journal of the American Oriental Society* 88 (1968) 104–108; J.-M. Durand *et al.*, 'Le combat du dieu de l'orage avec la Mer', *Mari: Annales de Recherches Interdisciplinaires* 7 (1993) 41–70 at 45; M. Bauks, *Die Welt am Anfang: Zum Verhältnis von Vorwelt und Weltentstehung in Gen 1 und in der altorientalischen Literatur*, Neukirchen-Vluyn 1997, 249–251.

[50] *Contra* Collard, *Euripides: Selected Fragmentary Plays*, vol. I, 269; M.L. West, '*Ab ovo*: Orpheus, Sanchuniathon, and the Origins of the Ionian World Model', *Classical Quarterly* 44 (1994) 289–307.

In fact, there can hardly be any doubt about its origin. It has recently become increasingly clear that Orphism not only borrowed from Oriental poems, but also was heavily influenced by Egyptian traditions. From the seventh century onwards, Lydian, Carian and Greek mercenaries had entered the service of the Pharaohs and even left their 'Kilroy was here' in Abu Simbel;[51] merchants traded in Naucratis in the Nile Delta,[52] and Egyptian religion must have gradually become better known ever since. In fact, it has already long been seen that the function of the Orphic Gold Leaves as 'passports', their dialogue form and their mention of fresh water, derive from the Egyptian *Book of the Dead*.[53] And indeed, Egyptian influence on Orphism has recently been stressed and documented by Burkert.[54] Now the separation of heaven and earth is a highly familiar motif in Egyptian religious literature and iconography. A Pyramid text (1208c) already speaks of the time when 'heaven was separated from earth, when the gods ascended to heaven'. The idea was taken up by Heliopolis and given its classic formulation: Shu separates the sky (Nut) from earth (Geb).[55] I therefore conclude that Orphism had taken this motif also from ancient Egypt.

The idea that the union of Heaven and Earth generates all living things does appear elsewhere in Greek tradition. We have already seen it in Aeschylus' *Danaids* (above), but it also occurs in fragments of

---

[51] See most recently P.W. Haider, 'Griechen im Vorderen Orient und in Ägypten bis ca. 590 v. Chr.', in: Ch. Ulf (ed.), *Wege zur Genese griechischer Identität*, Berlin 1996, 59–115 at 95–113; H. Hauben, 'Das Expeditionsheer Psamtiks II. in Abu Simbel (593/92 v. Chr.)', in: K. Geus and K. Zimmermann (eds), *Punica—Libyca—Ptolemaica: Festschrift Werner Huss*, Leuven 2001, 53–77; M. Bietak (ed.), *Archaische Griechische Tempel und Altägypten*, Vienna 2001; K. Kopanias, 'Der ägyptisierende "Branchide" aus Didyma', in: H. Klinkott (ed.), *Anatolien im Lichte kultureller Wechselwirkungen*, Tübingen 2001, 149–166; O. Carruba, 'Cario *Natri* ed egizio *n t r* "dio",' in: M. Fritz and S. Zeilfelder (eds), *Novalis Indogermanica: Festschrift für Günter Neumann zum 80. Geburtstag*, Graz 2002, 75–84.

[52] For Naucratis see most recently A. Müller, *Naukratis*, Oxford 2000; A. Bresson, *La cité marchande*, Bordeaux & Paris 2000, 13–63, 65–84 and 'Quatre emporia antiques: Abal, la Picola, Elizavetouskie, Naucratis', *Revue des Études Anciennes* 104 (2002) 475–505 at 496–505.

[53] F. Graf, *Eleusis und die orphische Dichtung Athens*, Berlin and New York 1974, 125–126; S. Morenz, *Religion und Geschichte des alten Ägypten*, Cologne 1975, 462–489; R. Merkelbach, 'Die goldenen Totenpässe: ägyptisch, orphisch, bakchisch', *Zeitschrift für Papyrologie und Epigraphik* 128 (1999) 1–13.

[54] Burkert, *Griechen und der Orient*, 79–106.

[55] For these and other texts see Morenz, *Ägyptische Religion*, 182–183; H. te Velde, 'The Theme of Separation of Heaven and Earth in Greek Mythology', *Studia Aegyptiaca* 3 (1977) 161–170.

Euripides where Sky (Aether) and Earth generate and recycle all life.[56] The striking aspect of our passage is that the human race is mentioned as well and, moreover, clearly as the most important 'product' of the cosmic union, since the passage is part of a speech in which Melanippe defends her infants. The mention of the human race, therefore, well fits the already noticed Orphic interest in anthropogony.

Burkert has posed the question whether cosmogonical poetry was sung during healing rituals, but admits that 'detailed documentation is still not available'.[57] He has not been able to adduce any specific Greek passage, and neither do we find it here. Yet it has been noted that the activities of Melanippe's mother Hippo display an 'exotic character',[58] since in addition to the cosmogony she also 'sang oracular songs to men, telling them cures and deliverances from their pains' (*Wise Melanippe* 16–17 Diggle, transl. Collard *et al.*). Now the 'exotic' usually derives from either certain traditional topoi or from reality.[59] And indeed, in our Greek texts there is one particular class of people associated with divination, the telling of a theogony and the healing of the sick, like those struck by epilepsy, viz. the Persian Magi.[60] Mention of them becomes increasingly frequent towards the end of the fifth century, and practising Magi have now turned up in Athens in the Derveni papyrus in a column (VI) that became known only in 1997.[61] Is it to be excluded that Euripides was thinking of them in particular, when he referred to the practices of Melanippe's mother?

---

[56] Euripides, frgs. 839, 898 (both re-edited by J. Diggle, *Tragicorum Graecorum fragmenta selecta*, Oxford 1998, 166–168), 1023 Nauck^2.

[57] Burkert, *Kleine Schriften*, vol. II, 64.

[58] Collard, *Euripides: Selected Fragmentary Plays*, 270.

[59] See, for example, J.A. González Alcantud, *La extraña seducción: Variaciones sobre el imaginario exótico de occidente*, Granada 1993.

[60] A. de Jong, *Traditions of the Magi*, Leiden 1997, 363 (theogony: Herodotus 1.132), 397–399 (divination). For the Magi at the time of Darius see now J. Kellens, 'L'idéologie religieuse des inscriptions achéménides', *Journal asiatique* 290 (2002) 417–464 at 448–457.

[61] As was simultaneously pointed out by Bremmer, 'The Birth of the Term "Magic"', in: Bremmer and J. Veenstra (eds), *The Metamorphosis of Magic*, Leuven 2002, 1–11 at 8–9 and Burkert, *Griechen und der Orient*, 126–132.

2.2. *Aristophanes*

After these serious cosmogonies, let us conclude with a brief look at a cosmogonical pastiche. In his *Birds* Aristophanes relates an ornithogony that displays now familiar but also new aspects of ancient Greek cosmogony:

> In the beginning there was Chaos and Night and Black Erebus and broad Tartarus, and there was no Earth or Aer or Heaven; and in the boundless recesses of Erebus, black-winged Night, first of all beings, brought forth a wind-egg, from which, as the seasons came round, there sprang Eros the much-desired, his back sparkling with golden wings, resembling the swift whirlings of wind. And he, mating by night with winged Chaos in broad Tartarus, produced as chicks our own race and first caused it to see the light. But of old there was no race of immortal gods, until Eros mixed all things together; then, as one thing mixed with another, Heaven came to be, and Okeanos and Earth, and all the imperishable race of blessed gods. Thus we are far older than all the blessed (693–703, transl. A. Sommerstein, slightly revised).[62]

The beginning of this cosmogony largely agrees with Hesiod but leaves out Earth, who was equally absent from the first generation in Acusilaus (*FGrH* 2 frg. 5 = frg. 6b–d Fowler), whom Aristophanes may also have known. It is rather surprising that Aristophanes continues by stating what was not yet there, such as Earth and Aer (the misty lower sky as opposed to Aither, the bright upper sky). The latter occurs in the Orphic-like cosmogony of pseudo-Epimenides (B 5 = frg. 6a,b), but does not seem to have been part of an early Orphic cosmogony. It was probably derived from Anaximenes, for whom Aer was the primal element.[63] In Aristophanes, Aer probably owes its mention to the fact that it is birds who are speaking. Having set the stage, Aristophanes now pulls a surprising rabbit from his hat. Night, whose role as 'first of all beings' we already discussed, laid a wind-egg. The term was customarily used for infertile eggs laid without preceding copulation (Dunbar

[62] For detailed discussions see now A. Pardini, 'L'*Ornitogonia* (Ar. *Av.* 693 sgg.) tra serio e faceto: premessa letteraria al suo studio storico-religioso', in: A. Masaracchia (ed.), *Orfeo e l'Orfismo*, Rome 1993, 53–65; N. Dunbar, *Aristophanes: Birds*, Oxford 1995, 437–444; A. Bernabé, 'Una cosmogonía cómica: Aristófanes, *Aves* 685ss.', in: J.A. López Férez (ed.), *De Homero a Libanio: Estudios actuales sobre textos griegos*, vol. II, Madrid 1995, 195–211.

[63] O. Kern, *De Orphei Epimenidis Pherecydis theogoniis quaestiones selectae*, Berlin 1888 (Diss.), 70; H. Demoulin, *Épiménide de Crète*, Brussels 1901, 122; Kirk, Raven and Schofield, *Presocratic Philosophers*, 144–147.

ad loc), and is probably used here to indicate the absence of a husband for Night.[64] Once again we seem to have here an Egyptian element, since in Egypt the egg assumes a cosmic significance. Admittedly, the egg is not attested in any other early Orphic text,[65] but it occurred both in the theogony of Pseudo-Epimenides (B 5 Diels-Kranz = frg. 6b Fowler), an author related to the Orphics, and in many later Orphic texts, such as the theogony of 'Hieronymus and Hellanicus' (frg. 54K = 79B). The egg may also have been mentioned in Aristophanes' comedy *Gerytades* (frg. 170 Kassel-Austin),[66] and it certainly appears in the Phoenician cosmogonies that so happily borrowed from the Orphics:[67] its earlier Orphic existence can therefore hardly be doubted.

From Hesiod came Eros, and his birth from an egg confirms the already observed lack of parents (§1). His prominent position here is probably indebted to his increasingly important role in Greek culture, as exemplified by Pherecydes (frgs. 72–73 Schibli), Parmenides (B 13 Diels-Kranz) and Empedocles (B 17 Diels-Kranz), whose thinking regularly approached that of the Orphics;[68] the Orphic tradition itself, as Pausanias could still observe among the Attic Lykomids (9.27.2);[69] poets like Sappho (frg. 198 Voigt), Aeschylus (frg. 44 Radt), and Euripides' *Hypsipyle* (1103 Bond = *OF* 2K = 65B: a passage with an Orphic colouring), and the mythographer Acusilaus (*FGrH* 2 frg. 5 = frg. 6a,b Fowler). An Orphic origin, then, seems not impossible. The theogonic function of Eros undoubtedly derives from its function in rites of maturation: 'the power of love, which maintained the social fabric of the civic community, likewise organised the ordering of things'.[70]

With this cosmic role of Eros, Aristophanes' description has come to a close; more would probably have been boring to his public. We may wonder, though, what actually would have been so funny about this passage. First, of course, the application of cosmogony to ornithogony. Second, the playing with the traditional elements and the adding of

[64] The other reasons adduced by Dunbar (*ad loc.*) seem a bit far-fetched.

[65] As is observed by J. Mansfeld, *Studies in Later Greek Philosophy*, London 1989, chap. XIV, 267, 291.

[66] Cf. A.C. Cassio, 'L'uovo orfico e il Geritade di Aristofane (frg. 164 K.)', *Rivista di Filologia e d'Istruzione Classica* 106 (1978) 28–31.

[67] West, '*Ab ovo*'. For a possible echo of the egg in Christian hagiography see P. Boulhol, *Analecta Bollandiana* 112 (1994) 282–284.

[68] C. Riedweg, 'Orphisches bei Empedokles', *Antike und Abendland* 41 (1995) 34–59.

[69] *OF* 28K = 20B; Euripides, *Hypsipyle* 57.23 Bond; Aristophanes, *Birds*, 700.

[70] J. Rudhardt, *Le role d'Eros et d'Aphrodite dans les cosmogonies grecques*, Paris 1986; C. Calame, *The Poetics of Eros in Ancient Greece*, Princeton 1999, 177–181 at 178 (quotation).

bird motifs, such as giving wings to Chaos. Third, we may wonder whether cosmogony as a genre did not always retain an air of strangeness for the average Greek. It was not supported by other traditions, and that is perhaps why Greek comedy made fun of ancient cosmogony, as we can also observe in Aristophanes' contemporary Cratinus (frg. 258 KA).

3. *Conclusion and final observations*

As we have seen in this contribution, the Greeks happily borrowed from the great poems of the Ancient Near East, just like the Israelite priests that composed the beginning of *Genesis*. Yet we cannot conclude in this manner, since such quotations and allusions across cultures should not lead us to overlook the fact that they function in wholly different cultural contexts.[71] Regarding that functioning I would like to conclude with three observations.

First, whereas the great Mesopotamian poems *Atrahasis*, *Enuma elish*, *Erra* and *Gilgamesh* tend to view the universe as created, except for the gods, Greek tradition looks at the universe as the fruit of a family tree. However, unlike Mesopotamia and Israel, Greece is hardly interested in the creation of man, which is absent from Homer and Hesiod's *Theogony*; moreover, if human creation is mentioned at all, males are born but only woman is created. In other words, even when taking over Near Eastern cosmogonies, the Greeks stuck to their old tradition about the genesis of man.

Secondly, the purpose of narrating the creation of the cosmos differs from culture to culture. In *Enuma elish* the poem's aim is to celebrate Marduk, whereas in *Genesis* the narrative hastily relates the creation in order to continue with the history of man. In Hesiod's *Theogony* the creation is the beginning of a much longer story about the rise to power of Zeus and the coming into being of his rule. In other words, even though Greece borrowed from the Near East, it used these cosmogonies rather differently, not only regarding their content but also their *Sitz im Leben*. Whereas in the Near East cosmogony was closely associated with ritual, Israel and Greece emancipated themselves in this

[71] As is well stressed by G.E.R. Lloyd, *Methods and Problems in Greek Science*, Cambridge 1991, 278–298; J. Haubold, 'Greek Epic: A Near Eastern Genre?', *Proceedings of the Cambridge Philological Society* 48 (2002) 1–19.

respect, which may well have enabled the turn towards philosophy that we witness in ancient Greece.

Thirdly, the cosmogonic accounts in *Genesis* and Hesiod are already much more rationalized than the Near Eastern ones. The author of *Genesis* provides us with a rather systematic account of the universe that is far from the 'just so' stories of *Atrahasis* or *Enuma elish*. Moreover, he relates the creation already in prose, just as the first Greek philosophers marked their new beginning by writing in prose.[72] This systematizing aspect is also true for Hesiod, if to a somewhat lesser account. We have recently become used to problematising the relationship between *mythos* and *logos*,[73] and undoubtedly Hesiod has to be located at the side of *mythos* (see also Tieleman, this volume, §2). Yet that does not mean that the philosophers were all on the side of *logos*. It would take some centuries before myth definitively lost out to philosophy, but in the increasingly reflective world of a literate Greece mythical cosmogonies could no longer satisfy the intellectual needs of this inquisitive culture.[74] Even if only hesitatingly, the first steps were now set on the centuries long path to modern investigations not so much of the creation but of the origin of heaven and earth and beyond.

APPENDIX: WHY AND WHEN WAS GENESIS 1:1 WRITTEN?

'In the beginning God created heaven and earth'. These proud and programmatic words of the first verse of the opening chapter of *Genesis* have become so familiar to us that we hardly realize how unusual they really are. Yet, as Hermann Gunkel (1862–1932) noted, 'Kein Wort gibt es in den Kosmogonien anderer Völker, das diesem ersten Wort

[72] A. Laks, 'Écriture, prose, et les débuts de la philosophie grecque', *Methodos* 1 (2001) 131–151 (I owe a copy of this article to the kindness of André Laks); C. Kahn, 'Writing Philosophy: Prose and Poetry from Thales to Plato', in: H. Yunis (ed.), *Written Texts and the Rise of Literate Culture in Ancient Greece*, Cambridge 2003, 139–161. In spite of the title, S. Goldhill, *The Invention of Greek Prose*, Oxford 2002 does not discuss the invention of prose.

[73] See most recently R. Buxton (ed.), *From Myth to Reason?*, Oxford 1998; B. Lincoln, *Theorizing Myth*, Chicago & London 1999, 3–18 ('The Prehistory of *Mythos* and *Logos*'); K. Morgan, *Myth and Philosophy from the Presocratics to Plato*, Cambridge 2000, 15–45; F. Graf, 'La génèse de la notion de mythe', in: J.A. López Férez (ed.), *Mitos en la literatura griega arcaica y clásica*, Madrid 2002, 1–15 at 2–6.

[74] For the importance of writing for the development of philosophy see now the stimulating observations of M.M. Sassi, 'La naissance de la philosophie de l'esprit de la tradition', in: Laks and Louguet, *Qu'est-ce que la Philosophie Présocratique?*, 55–81.

der Bibel gleichkäme'.[75] And indeed, none of the great Mesopotamian and Anatolian civilizations that were Israel's neighbours would have recognized itself in these words, as a creator 'of heaven and earth' only occasionally occurs in Akkadian and Assyrian texts,[76] and not at all in the cosmogonies of the Sumerians, Hittites and Phoenicians. Only in Egypt it was believed that the god Ptah created by his word everything that exists today.[77] But not even accounts of Egyptian cosmogony provide the same epigrammatic beginning as the Israelite text. It is therefore not surprising that the standard commentaries on *Genesis* find it difficult to provide a satisfactory explanation for its origin and position in the text.[78] It has even been suggested that, at some stage, the first verse was added to a pre-existing account;[79] in any case, the exact translation remains debated.[80] Recent studies of the beginning of *Genesis* have looked for help to the Mesopotamian world with its great epics of creation,[81] to the immediate North (the Phoenician world)[82] and to the South, the Egyptians,[83] but none of these cultures provides a proper parallel for Gen 1:1. As the final redaction of *Genesis* is now generally dated to the Achaemenid period, it is rather surprising to note that no Old Testament scholar seems to have looked to the Persians for an answer.

---

75 H. Gunkel, *Die Urgeschichte und die Patriarchen*, Göttingen 1911, 101; W.H. Schmidt, *Die Schöpfungsgeschichte der Priesterschrift*, Neukirchen-Vluyn 1973^{3}, 75: 'fehlt für den sich sprachlich scharf abhebenden V 1 ein Vorbild'.

76 *Enuma elish* VII.86, where Marduk is called 'Mummu, fashioner of heaven and earth'; W. von Soden, *Sumerische und Akkadische Hymnen und Gebete*, Stuttgart 1953, 321 no. 56.9, where Shamash is called 'der Schöpfer von allem und jedwedem im Himmel und auf der Erde'; G. Frame, *Rulers of Babylonia*, Toronto 1995, 197, where Marduk is called 'creator of heaven and netherworld' (time of Ashurbanipal). For other, less closely resembling passages see *Chicago Assyrian Dictorionary* B 88b (2'), K 504b, M/2 197b (1.a).

77 See K. Koch, 'Wort und Einheit des Schöpfergottes in Memphis und Jerusalem', *Zeitschrift für Theologie und Kirche* 62 (1965) 251–293, reprinted in his *Studien zur alttestamentlichen und altorientalischen Religionsgeschichte*, Göttingen 1988, 61–105; V. Notter, *Biblischer Schöpfungsbericht und ägyptische Schöpfungsmythen*, Stuttgart 1974, 23–26; J.P. Allen, *Genesis in Egypt*, New Haven 1988, 38–47.

78 Cf. C. Westermann, *Genesis*, vol. I, Neukirchen-Vluyn 1974, 130–141; H. Seebass, *Genesis*, vol. I, Neukirchen-Vluyn 1996, 65; M. Baasten, 'Beginnen bij het begin—Over Genesis 1:1', *Alef Beet* 12/1 (2002) 13–26.

79 Schmidt, *Schöpfungsgeschichte*, 74–75; Bauks, *Die Welt am Anfang*, 91.

80 See most recently M. Weippert, 'Schöpfung am Anfang oder Anfang der Schöpfung? Noch einmal zu Syntax und Semantik von Gen 1,1–3', *Theologische Zeitschrift* 60 (2004) 5–22.

81 Bauks, *Die Welt am Anfang*, 230–267.

82 H. Niehr, *Der höchste Gott*, Berlin 1990.

83 Bauks, *Die Welt am Anfang*, 147–230.

This does not mean that nobody ever noticed Persian influence regarding Israel's ideas about the creation. Exactly forty years ago from the time of my writing, Morton Smith suggested that many themes in *Deutero-Isaiah* 40–48 depend on Cyrus' proclamation concerning his conquest of Babylon. The similarities pointed out by Smith are certainly there, if perhaps less in number than he suggests, but, more importantly, Smith also argued that the prominence of the theme of Yahweh's creation of the world in these very chapters depended on Persian cosmological material.[84] To prove his point, he compared *Yasna* 44, a series of questions addressed to Ahuramazda, and he concluded that the author of *Deutero-Isaiah* 40–48 had derived its cosmology from the Persians.[85]

It is indeed striking that the combination of *bara* ('to create') + *shamayim* ('heaven') + *erets* ('earth') in one verse occurs especially in the chapters identified by Smith. In addition to Gen 1:1, Gen 2:4 and Deut 4:32 (same words, but very different combination), the combination occurs only in Isa 42:5 ('Thus says God, the LORD, who created the heavens and stretched them out, who spread out the earth and what comes from it, who gives breath to the people upon it and spirit to those who walk in it'); Isa 45:8 ('Shower, O heavens, from above, and let the skies rain down righteousness; let the earth open, that salvation may spring up, and let it cause righteousness to sprout up also; I the LORD have created it'); Isa 45:12 ('I made the earth, and created humankind upon it; it was my hands that stretched out the heavens, and I commanded all their host'); Isa 45:18 ('For thus says the LORD, who created the heavens [he is God!], who formed the earth and made it [he established it; he did not create it a chaos, he formed it to be inhabited!]: I am the LORD, and there is no other') and Isa 65:17 ('For I am about to create new heavens and a new earth; the former things shall not be remembered or come to mind', all translations from the NRSV).[86]

---

[84] For Ahuramazda as creator god see G. Ahn, 'Schöpfergott und Monotheismus. Systematische Implikationen in der neueren Gatha-Exegese', in: M. Dietrich and I. Kottsieper (eds), *'Und Mose schrieb dieses Lied auf': Studien zum Alten Testament und zum Alten Orient. Festschrift für Oswald Loretz*, Münster 1998, 15–26.

[85] M. Smith, *Studies in the Cult of Yahweh*, ed. S.J.D. Cohen, 2 vols, Leiden 1996, vol. I, 73–83 ('II Isaiah and the Persians', 1963[1]). For Smith (1915–1991) see W.M. Calder III, 'Morton Smith', *Gnomon* 64 (1992) 382–384; S.J.D. Cohen, 'In Memoriam Morton Smith', in: Smith, *Studies*, vol. II, 279–285.

[86] But note also Isa 44:24: 'I am the LORD, who made (the verb used is *'asah* not *bara*) all things, who alone stretched out the heavens, who by myself spread out the earth.'

But do these verses of *Deutero-Isaiah* also point to a Persian influence? As Smith was not an expert *in rebus Persicis*, he consulted his friend and colleague Elias Bickerman, who had long been interested in the history of Israel during the Persian period.[87] Bickerman pointed Smith to an inscription of Xerxes (485–465), found in Persepolis in Old Persian (two copies), Elamite and Babylonian, but since also found in Pasargadae in 1963. At its beginning the Persian king proclaims: 'A great god (is) Ahuramazda, who created this earth, who created yonder heaven, who created man, who created blissful happiness for man, who made Xerxes king, the one king of many, the one master of many'.[88] Bickerman knew this inscription (XPh) from the *ANET* (316–317), which uses a translation by Ernst Herzfeld (1879–1948),[89] and he concluded that 'II Isaiah's insistence that Yahweh is the creator might thus be seen as reaction, but, reaction or not, its form and presumably its content have been shaped by Persian tradition'.[90]

Bickerman's conclusion seems reasonable, but he does not explain the striking position of the idea of God as creator at the very beginning of *Genesis*. In fact, neither he nor Smith considered Gen 1:1 in this context. However, in the course of a discussion of Assyrian and early Greek cosmologies, Walter Burkert noted in passing: 'Nach einer Inschrift des Darius aus Persepolis ist es Ahura Mazda, der "Himmel und Erde geschaffen hat" (F.H. Weissbach, *Die Keilinschriften der Achämeniden*, Leipzig 1911, 85), ganz wie Jahwe Gen 1,1'.[91] We can now combine the insights of Bickerman and Burkert by looking not so much to the tomb of Darius at Naqsh-I Rustam (DNa) or Persepolis, where the theme of Ahuramazda as creator occurs only in a few inscriptions, but to a different Persian capital, namely Susa.

After Darius I (521–486) had built his palace in Susa, he recorded the building process in a trilingual inscription (Old Persian, Elamite

---

[87] See, for example, his *From Ezra to the Last of the Maccabees*, New York 1962; *Four Strange Books of the Bible*, New York 1967; *Studies in Jewish and Christian History*, vol. I, Leiden 1976, 72–108 ('The Edict of Cyrus in Ezra I', 1946[1]). For Bickerman (1897–1981) see A. Momigliano, *Settimo contributo alla storia degli studi classici e del mondo antico*, Rome 1984, 371–375; F. Parente, 'Ricordo di Elias Joseph Bickerman', *Athenaeum* 60 (1982) 237–244; C. Bonnet and A. Marcone, '"Mon activité académique ici est fini." Due lettere inedite di Elias Bickerman à Franz Cumont', *Rivista Storica Italiana* 94 (2002) 239–245.

[88] For text, translation and commentary see now R. Schmitt, *The Old Persian Inscriptions of Naqsh-I Rustam and Persepolis*, London 2000, 88–95.

[89] E. Herzfeld, *Altpersische Inschriften*, Berlin 1938, 27–35.

[90] Bickerman *apud* Smith, *Studies*, 82–83.

[91] Burkert, *Kleine Schriften*, vol. II, 229 note 32.

and Babylonian) on clay tablets, glazed bricks and marble tables, which he put up all over Susa (DSf). So far about thirteen Old Persian, twelve Elamite and twenty-seven Babylonian copies or fragments of this inscription, called 'Charte du Fondation du Palais' by its *editio princeps*,[92] have been found. Its actual beginning recurs on another, equally frequently displayed inscription in Susa (DSe) with Darius' enumeration of the peoples dominated by him, of which eleven Old Persian, five Elamite and three Babylonian copies or fragments have been found. In other words, the text of the beginning of these inscriptions must have been visible all over Susa. It therefore cannot have escaped foreign visitors from, say, Ionia or Israel. They would have seen an inscription that began as follows: 'A great god (is) Ahuramazda, who created this earth, who created yonder heaven, who created man, who created blissful happiness for man, who made Darius king, the one king of many, the one master of many' (§1, transl. Schmitt).[93] Darius, then, prefaced his accounts with a cosmogony in which he stressed the creation of heaven and earth by his favourite god, and, as can now be seen, Xerxes followed his father in this tradition.[94]

Unfortunately, we cannot be certain about the exact date of the building of Darius' palace in Susa. It seems to have started about 520 BC, around the same time as the building in Persepolis.[95] The Persepolis Old Persian versions, though, do not yet mention the creation of heaven and earth by Ahuramazda.[96] This shows that the standard beginning formula had not yet developed at that time and must be

---

[92] V. Scheil, *Inscriptions des Achémenides à Suse*, vol. I, Paris 1929, 3–34 and *Actes juridiques susiens (suite)—Inscriptions des Achémenides à Suse (supplément et suite)*, Paris 1933, 105–115. For the most recent editions of this inscription see F. Vallat, 'Deux inscriptions élamites de Dareios Ier (DSf et DSz) (1)', *Studia Iranica* 1 (1972) 3–13; M.-J. Steve, *Village Royale de Suse*, vol. VII, Paris 1987, 64–77 (Old Persian and Akkadian). For the most recent studies see H. Klinkott, 'Die Funktion des Apadana am Beispiel der Gründungsurkunde von Susa', in: M. Schuol *et al.* (eds), *Grenzüberschreitungen*, Stuttgart 2002, 235–257; W. Henkelman, '"Dit paleis dat ik in Susa bouwde." Bouwinscriptie(s) van koning Dareios I', in: R. Demarée and K. Veenhof (eds), *Zij schreven geschiedenis. Historische documenten uit het Oude Nabije Oosten (2500–100 v. Chr.)*, Louvain 2003, 372–386.

[93] For a translation that also records the differences between the three versions, see P. Lecoq, *Les inscriptions de la Perse achéménide*, Paris 1997, 234–237: 'Ahuramazda est le grand dieu qui a créé cette terre ici, qui a créé ce ciel là-bas, qui a créé l'homme, qui a créé le bonheur pour l'homme, qui a fait Darius roi, unique roi de nombreux, unique souverain de nombreux'.

[94] For Xerxes see also the Persepolis inscriptions XPa, XPb, XPc, XPd and XPh.

[95] P. Briant, *Histoire de l'empire perse de Cyrus à Alexandre*, 2 vols, Paris 1996, vol. I, 177–180 (building of Susa), 180–182 (building of Persepolis); vol. II, 934 (dates).

[96] Schmitt, *Old Persian Inscriptions*, 56.

dated to about 515–510 BC. Now neither Darius' father Cyrus nor his son Xerxes had the same relationship with Ahuramazda as Darius himself. The latter's preference is well illustrated by the fact that Ahuramazda is mentioned 63 times in his famous inscription of Behistun (DB), whereas all the other gods are mentioned only once; similarly, it is Ahuramazda who is incessantly invoked in Darius' prayers. One cannot speak of monotheism in this case, but Darius evidently associated his own rise to power with a hegemonic position within the pantheon for Ahuramazda.[97]

The philosopher Heraclitus from Ephesus, who was a contemporary of Darius and had almost certainly met Persian Magi in his home town,[98] seems to have already reacted to this new doctrine of a creator god of heaven and earth, although not with approval: 'This world order ... no one of the gods or men has made' (B 30 Diels-Kranz = 51 Marcovich).[99]

Other reactions to Darius' claims may well have been the beginning of *Genesis* and parts of *Deutero-Isaiah*, the more so since it is precisely in the latter that Jahweh is elaborately hailed as incomparable (Isa 40:12–31, 46:5–13) and unique (Isa 43:8–13, 44:6–8, 45:18–25). Unfortunately, neither treatise can be accurately dated. However, the present text of *Genesis* must postdate the so-called *Priesterschrift*, the generally accepted source for the first chapter of *Genesis*, which is commonly dated to the period 550–490 BC.[100] As regards *Deutero-Isaiah*, the text mentions Cyrus (Isa 41:1–7, 42:5–9, 44:24–28, 45:1–8) and anticipates the fall of Babylon (539 BC: Isa 43:14–15, 47:1–15, 52:11–12), but it also alludes to the rebuilding of the temple in Jerusalem, which was completed in 515 BC (Isa 44:28). Whereas the early studies dated *Deutero-Isaiah* to

[97] For the close tie between Darius and Ahuramazda, whom Darius promoted to the most prominent position in the Persian pantheon, see Briant, *Histoire de l'empire perse*, vol. I, 138–139.

[98] For his mention of Magi see Heraclitus B 14 Diels-Kranz = 87 Marcovich, cf. Burkert, *Die Griechen und der Orient*, 156 note 28, who reads *nyktipolois magois*. However, with F. Graf, *Magic in the Ancient World*, Cambridge, MA 1997, 21, I read: *nyktipolois*: *magois*, etc. For the various opinions about the fragment see Bremmer, 'The Birth of the Term "Magic"', 2 note 10.

[99] As noted by Burkert, *Kleine Schriften*, vol. II, 229.

[100] Source: E. Zenger *et al.*, *Einleitung in das Alte Testament*, Stuttgart 1998³, 148: 'Dass der Anfang von P^g (Priesterschrift) in Gen 1 vorliegt, ist unbestritten'. Date: H. Seebass, 'Pentateuch', in: *Theologische Realenzyklopädie* (*TRE*), vol. 26 (1996) 185–209 at 192: 'Zwischen ca. 550 und den Anfang des 5. Jhu. v. Chr.'; E. Zenger, 'Priesterschrift', in: *TRE*, vol. 27 (1997) 435–446 at 439: 'eine Datierung um 520 v. Chr.'.

the years immediately preceding the fall of Babylon, the most recent syntheses agree that the text is a composition of chronologically heterogeneous materials which do not allow a precise dating.[101]

Given these close similarities between the Persian and Israelite texts, it is hard to believe that the authors of Gen 1:1 and the relevant chapters of *Deutero-Isaiah* did not, directly or indirectly, observe Ahuramazda's rise to prominence under Darius; they will also have observed Darius' claim that he was the creator of heaven and earth. Both authors (or their sources) may well have seen or heard of the Babylonian versions that reverse the Old Persian order and read 'who has created heaven, who has created this earth' (Susa: DSf) or, even closer to the text of *Genesis*, 'who created heaven and earth' (Persepolis: DPg).[102] Apparently, they did not want to pass over this claim for Ahuramazda as the creator and wrote a competing claim for Jahweh as the creator of heaven and earth. We might even speculate that Gen 1:1 was prefixed to the creation account of the *Priesterschrift* in the early years of Darius' rule, perhaps for the occasion of the completion of the temple in 515 BC. However, all speculation is futile, since we lack stable chronological anchors regarding the development and transmission of the texts of *Genesis* and *Deutero-Isaiah*. What I hope to have established, though, is that future analyses of Israel's creation accounts can only neglect the Persian evidence at their peril.[103]

---

[101] D. Michel, 'Deuterojesaja', in: *TRE* 8 (1981) 510–530; H.-J. Hermisson, 'Deuterojesaja', in: *Religion in Geschichte und Gegenwart*[4], vol. 2 (1999) 684–688.

[102] For the position of the Babylonian language in Darius' time see J. Oelsner, 'Babylonische Kultur nach dem Ende des babylonischen Staates', in: R.G. Kratz (ed.), *Religion und Religionskontakte im Zeitalter der Achämeniden*, Gütersloh 2002, 49–73 at 58–61.

[103] I am most grateful to Marten Stol, Eibert Tigchelaar and Bob Fowler for comments and information. The latter also gracefully let me use his own forthcoming commentary on the relevant sections of the Greek mythographers. The Appendix has much profited from a discussion in the Groninger Oudtestamentische Kring. Richard Buxton kindly and skilfully corrected my English.

COSMIC GODS AND PRIMORDIAL CHAOS IN HELLENISTIC AND ROMAN PHILOSOPHY: THE CONTEXT OF PHILO'S INTERPRETATION OF PLATO'S *TIMAEUS* AND THE BOOK OF GENESIS

John Dillon

Introduction

In view of the particular perspective of this conference, it seems appropriate to me to approach the theme that I have been assigned from the point of view of a thinker who, while thoroughly steeped in the biblical and wider Jewish tradition, is yet thoroughly alert to the latest tendencies in Hellenic philosophy, and that is Philo of Alexandria.

In the work with which he inaugurates his exposition of the Jewish Laws, the *De opificio mundi*, Philo, as we know, expounds the higher significance of Moses' account of divine cosmogony in the first chapter of *Genesis* with a constant eye on the *Timaeus* of Plato. Such a statement is no longer news, especially after the magisterial investigations of David Runia;[1] what still does merit some discussion, however, is precisely what interpretation of the *Timaeus* Philo is working with, and the answer to that is not simple at all. It is the investigation of this question which will lead us, I hope, to a more accurate view of how the relations between an active, or demiurgic, principle and a passive, primordially chaotic, material principle were understood in the later Hellenistic and early Roman Imperial period.

[1] In his monograph *Philo of Alexandria and the* Timaeus *of Plato*, Leiden 1986, and, more recently, in his contribution to the Philo of Alexandria Commentary series, *Philo of Alexandria: On the Creation of the Cosmos according to Moses; Introduction, Translation and Commentary*, Leiden 2001

1. *Plato and Aristotle on (pre-)cosmic chaos*

All later Platonist speculations about the creation of the world take their start from certain key passages of Plato's *Timaeus*, so we may take our start from a consideration of them. First, *Timaeus* 28B:[2]

> So concerning the whole heaven or world (οὐρανὸς ἢ κόσμος)—let us call it by whatsoever name may be most acceptable to it—we must ask the question which, it is agreed, must be asked at the outset of inquiry concerning anything: has it always been in existence, without any source of becoming; or has it come to be, starting from some beginning (ἀπ' ἀρχῆς τινος ἀρξάμενος)? It has come to be (γέγονεν).

Here the crucial word is γέγονεν, 'it has come to be'. Ever since Aristotle launched a criticism of Plato, presumably initially in oral disputations, ridiculing the postulate that something which had a beginning could be deemed to have no end, it had been a concern of members of the Academy, beginning with Plato's former associates, and immediate successors, Speusippus and Xenocrates, to deny that Plato meant the *Timaeus* creation account to be taken literally (see also Tieleman, this volume, §2). Their formula seems to have been that Plato presented it in this form simply 'for purposes of instruction' (διδασκαλίας χάριν).[3] This position in turn comes under attack from Aristotle, at *De caelo* 279b32ff.:[4]

> The defence of their own position attempted by some of those who hold that the world is indestructible but generated, has no validity. They claim that what they say about the generation of the world is analogous to the diagrams drawn by mathematicians: their exposition does not mean that the world ever was generated, but is used for instructional purposes (διδασκαλίας χάριν), since it makes things easier to understand, just as the diagram does for those who can watch it being constructed.
>
> But the analogy, as I say, is a false one. In the construction of geometrical figures, when all the constituents have been put together, the resulting figure does not differ from them; but in the expositions of these philosophers the result is not the same as the components, but rather produces an impossible situation; for the earlier and later assumptions are contra-

[2] I borrow here the translation of F.M. Cornford, with minor alterations.

[3] We are so informed by a scholiast on *De caelo* 279b32ff. (= Speusippus, frg. 61a, ed. L. Tarán), who tells us that 'Xenocrates and Speusippus, in an attempt to support Plato, declared that Plato had not held that the cosmos was generated (γενητός), but ungenerated, claiming that it was generated for the sake of instruction, in order to make the process more easily grasped and perspicuous.'

[4] I borrow here the Loeb translation of W.K.C. Guthrie, with minor alterations.

> dictory. They say that order arose from disorder,[5] but a thing cannot be at the same time in order and in disorder. The two must be separated by a process of generation involving time. In geometrical figures, on the other hand, there is no separation by time.

Here Aristotle makes an effort to undermine the geometrical analogy proposed by his former colleagues in the Academy. Three straight lines, for example, he would argue, when taken separately, are not in their nature antithetical to the existence of the triangle for the formation of which they are combined; but the pre-cosmic chaos described by Plato, which his Demiurge has to take in hand and bring to order, is the very antithesis of that order.

This argument is ingenious, but not, I think, compelling. The point that the Platonists would make is that there never *was* a pre-cosmic chaos, so that all that Plato is describing is a feature of the world as it now is, which is an irreducible element of disorder and imperfection that is inseparable from the formation of a physical realm, and which the creative World Soul, or Cosmic Intellect—or whatever we want to make of the demythologized Demiurge—cannot entirely eliminate.

These, at any rate, were the first shots in a very long campaign, the reverberations of which certainly reached the ears of Philo in first-century BCE Alexandria. Before going any further, however, let us look also at *Timaeus* 30A, to which Aristotle has alluded above:

> Desiring, then, that all things should be good and, so far as might be, nothing imperfect, God took over all that is visible—*not at rest, but in discordant and unordered motion* (οὐχ ἡσυχίαν ἄγον ἀλλὰ κινούμενον πλημμελῶς καὶ ἀτάκτως)—and brought it from disorder into order, deeming that the former state was in every way better than the latter.

On the literal level, we are certainly here presented with the scenario of a pre-cosmic chaos, and this scenario is returned to later in the work, at 53A, where we find a description of the effect on what is now called the 'Nurse of what comes to be' (γενέσεως τιθήνη), or 'Receptacle' (ὑποδοχή, πανδεχές), of the imposition of the basic triangles and geometrical figures, which are the mode in which the Forms are projected upon it by the Demiurge in the mythical account:

> In this same way (sc. as corn is shaken about in winnowing-baskets) at that time the four kinds (i.e. the elements) were shaken by the Receptacle, which itself was in motion like an instrument for shaking, and it separated the most unlike kinds farthest apart from one another, and

[5] A reference to *Timaeus* 30A, to be quoted presently.

> thrust the most like closest together; whereby the different kinds came to have different regions, *even before the ordered whole consisting of all of them came to be*. Before that, all three kinds were without proportion or measure. Fire, water, earth, and air possessed, indeed, some vestiges (ἴχνη) of their own nature, but were altogether in such a condition as we should expect for anything when God is absent from it.

Here we have a vivid picture of a pre-cosmic chaos being brought to order by the action of a Creator. However, I am on record elsewhere[6] as suggesting that there is much that is odd, and even incoherent, about this description, which led me then to conclude that Plato means us to deduce that it is not to be taken literally, and I would hold to that today, despite some vigorous disputation with colleagues in the interval. What Plato here presents us with, after all, is something of a contradiction in terms. We have previously learned that what we thought of as the four basic elements are really composed of combinations of triangles (forming four of the five basic Platonic bodies), which derive from the forms in the Paradigm; but now we seem to see some kind of 'traces' (ἴχνη) or prefigurations of these already in the Receptacle, sloshing around in a random and chaotic way—but yet, it would seem, also beginning to sort themselves out into heavier and lighter (though not in a way that will ever come to anything). But what could these 'traces' possibly be? And how could they *begin* to sort themselves out, without such a process developing to some conclusion?

My solution is that Plato means us to ask ourselves these questions, and to conclude (as did his immediate followers) that there was never a pre-cosmic stage in the world's creation, but rather that the material substratum, by its very nature, produces a certain degree of distortion in the combination of the elemental bodies which the cosmic Intellect, despite its creative power and benevolence, cannot entirely overcome, and that is what produces our imperfect world.

If all that is so, however, we have to find some other acceptable meaning for the apparently blunt and uncompromising γέγονεν of *Timaeus* 28B; and in fact this is a task to which Platonists in the generations after Plato's death did turn themselves. Plainly, if γέγονεν does not have its normal meaning, it must have some other, more rarefied, one. Much later, in the mid-second century CE, we find evidence, from the Pla-

[6] 'The Riddle of the *Timaeus*: Is Plato Sowing Clues?', in: M. Joyal (ed.), *Studies in Plato and the Platonic Tradition: Essays Presented to John Whittaker*, Aldershot 1997, 33–37.

tonist Calvenus Taurus, of a set of fully *four* non-literal meanings that had been attached by his time to this word,[7] but we have evidence that such speculation goes back to the Old Academy. The first man attested[8] to have composed a commentary (of whatever degree of comprehensiveness) on Plato is Crantor of Soli, the associate of Polemon, the third head of the Academy, in the early years of the third century BCE. On this question, Proclus (*In Platonis Timaeum commentaria* I.27.8–10, ed. E. Diehl) gives us the following information:

> Commentators on Plato such as Crantor[9] declare that the cosmos is 'generated' (γενητός) in the sense of being produced by a cause other than itself (ὡς ἀπ' αἰτίας ἄλλης παραγόμενον), and not self-generating nor self-substantiating.

We may conclude from this, I think, that there was already in the Academy in Crantor's time a fairly thorough-going exegesis of the *Timaeus* based on the assumption of a non-literal interpretation of the creation account. What this then leaves us with, in effect, is an active cause, which may be interpreted as a demiurgic Intellect, or rational World-Soul, and a passive, 'material' principle, which is eternally being moulded and ordered by the active principle. The product of this process, the physical world, is *genêtos* only in the sense of being the eternal consequence of this process, as opposed to arising by some spontaneous activity, such as the whirl and progressive linking together of atoms—as in the Democritean scenario which Plato so disliked.

We find confirmation that this was how Plato's doctrine was viewed, not only from within the school, from a doxographical report relayed by Cicero,[10] which David Sedley[11] has persuasively argued to emanate

---

[7] For a discussion of these, see my *The Middle Platonists*, London 1996 (Revised ed. with a new afterword; 1977[1]), 242–244.

[8] By Proclus (*In Platonis Timaeum commentaria* I.76.1–2, ed. E. Diehl). Proclus is a late and not entirely reliable authority, but he cannot have had *no* reason for making this assertion, and we do in fact have a number of quite detailed comments by Crantor on the *Timaeus* preserved by other sources, notably Plutarch.

[9] Proclus here employs the formulation *hoi peri X*, 'those about X', but this very often means no more than the man himself, so that we do not need to postulate a 'school' of Crantorians.

[10] In *Academica* I.24–29, where M. Terentius Varro is presented as setting out the doctrines of the 'Old Academy'.

[11] In his important contribution to the 2000 Symposium Hellenisticum in Lille, 'The Origins of Stoic God', in: D. Frede and A. Laks (eds), *Traditions of Theology: Studies in Hellenistic Theology*, Leiden 2002, 41–83.

from Polemon, but also from no less an authority than Theophrastus, in his *Epitomes of Physical Doctrines*. First Polemon, or 'the school of Polemon':

> The topic of Nature, which they treated next, they approached by dividing it into two principles, the one the creative (*efficiens* = ποιητική), the other at this one's disposal, as it were, out of which something might be created. In the creative one they deemed that there inhered power (*vis* = δύναμις), in the one acted upon, a sort of 'matter' (*materia* = ὕλη); yet they held that each of the two inhered in the other, for neither would matter have been able to cohere if it were not held together by any power, nor yet would power without some matter (for nothing exists without being necessarily somewhere).[12] But that which was the product of both they called 'body' (*corpus* = σῶμα) and, so to speak, a sort of quality.[13]

What we have here, then, is a doxographical version of what would seem to be the later Academic version of the doctrine of the *Timaeus*, suitably demythologized. That this was accepted even in the Peripatos as being Plato's true doctrine is borne witness to most interestingly by no less an authority than Theophrastus, in the work mentioned above:[14]

> After these (sc. the earlier philosophers of nature) came Plato, prior to them in reputation and ability, though later than them in date. He concerned himself chiefly with first philosophy (sc. metaphysics), but also attended to the visible world (τὰ φαινόμενα), taking up the enquiry concerning nature. Here he wished to make the principles two in number, one underlying things as matter, which he calls the 'all-receptive' (πανδεχές), the other being the cause and source of movement, and this he attaches to the power of God and of the Good.

This is a most interesting little passage, for the assumptions that it makes. First, we may note that the material principle is identified with the Receptacle of the *Timaeus*, here given one of the titles that

---

[12] This seems to be a deliberate reminiscence of a passage from the *Timaeus*, 52C: 'Everything that exists must necessarily be in some place (ἔν τινι τόπῳ)'—pointing to the origin of the doctrine contained in this passage.

[13] Cicero is apologising here for coining the neologism *qualitas* to render the Greek ποιότης, in its turn a neologism of Plato in the *Theaetetus*, 182A, coined to describe what it is that an active principle (ποιοῦν) brings about in a passive principle—in this case a sense-organ. However, it would seem that, in the later Academy, helped by a perceived etymological connexion between ποιοῦν and ποιόν or ποιότης, this term became generalized as a description of forms in matter; and that is how it is being used here.

[14] Quoted by Simplicius, *In Aristotelis physicorum libros commentaria* IX, p. 26.5–15, ed. H. Diels, *Commentaria in Aristotelem Graeca* = Theophrastus, frg. 230, ed. W.W. Fortenbaugh *et al.*

it is accorded in *Timaeus* 51B;[15] then, that the Demiurge seems to be identified with the Good of the *Republic*, as a first principle of all things. This would imply, it seems to me, that the Paradigm, which in the myth of the *Timaeus* is presented as being external to the Demiurge, and, if anything, prior to him, must in fact be simply the contents of his intellect. It is further assumed here, it seems to me, that the process of interaction between the active and passive principles is eternal; there is no question of the creation of the world in time.

2. *Philo's interpretation of the 'Timaeus'*

It is such a scenario as this that Zeno of Citium and the Stoic School inherited from the Academy of Polemon. They gave it a further materialist, immanentist twist, but it is in turn a de-materialized, transcendentalized, and Pythagoreanized version of this that descends to Philo of Alexandria, through, perhaps, the mediation of some such figure as his fellow-Alexandrian and older contemporary Eudorus.[16] In his *De opificio mundi*, as I stated at the outset, he is concerned to apply this version of the *Timaeus* to the creation account at the beginning of *Genesis*.

The LXX version, which is what Philo is working with, does, as you recall, present a rendering of the original *we ha-aretz haytah tohu wa bohu* as ἡ δὲ γῆ ἦν ἀόρατος καὶ ἀκατασκεύαστος (Gen 1:2a), 'and the earth was invisible and unstructured', a form of expression which lends itself more readily to philosophical reinterpretation (see also Noort, this volume, §1, and Van Kooten, §1.2a). What Philo wants to make of this is in fact thoroughly philosophical, though he is certainly much concerned to stick closely to the text of Moses. Above all, he is opposed to the theory (which is that of Aristotle and the Peripatetic School) that the cosmos is eternal and uncreated. This, for him, sets up the cosmos as co-ordinate with God, and seriously derogates from God's omnipotence and providential control of it (see also Van den Berg, this volume, §4). He lays out his position initially in §§7–9:[17]

---

15 Aristotle too, we may note, has no hesitation about identifying the Receptacle of the *Timaeus* with his own concept of Matter, e.g. at *Physica* IV.2, 209b33ff.

16 On Eudorus, see my account in *The Middle Platonists*, 114–135. We cannot, however, it must be said, put our finger with certainty on any point of contact between Eudorus and Philo.

17 I borrow here the translation of David Runia, *Philo of Alexandria: On the Creation of the Cosmos*, with some modifications.

> There are some people who, having more admiration for the cosmos that for its maker, declared the former both ungenerated and eternal (ἀγένητός τε καὶ ἀίδιος), while falsely and impurely attributing to God much idleness. What they should have done was the opposite, namely be astounded at God's powers as Maker and Father,[18] and not show more reverence for the cosmos than is its due.
>
> Moses, however, had not only reached the summit of philosophy, but had also been instructed in the many and most essential doctrines of nature by means of oracles (χρησμοί).[19] He recognised that it is absolutely necessary that among existing things there is on the one hand an active cause (δραστήριον αἴτιον), and on the other a passive element (παθητόν), and that the active cause is the absolutely pure and unadulterated intellect of the universe, superior to moral excellence (ἀρετή), superior to knowledge, and even superior to the good and the beautiful itself.[20] But the passive element, which of itself is without soul and unmoved,[21] when set in motion and shaped and ensouled by the intellect, changed into the most perfect piece of work, this cosmos.

This serves to situate Philo interestingly within the philosophical milieu of his time. He is well aware of the controversy over the eternity or otherwise of the world,[22] and he knows what he is opposed to. But what position precisely does he take up? Obviously, for him, the world has to be in some sense 'created', since only that is consistent with God's omnipotence. But in what sense? Simple creation out of nothing at a point in time is not, I think, an option for anyone trained in Greek

[18] A reminiscence, of course, of Plato's terminology at *Timaeus* 28C3.

[19] Presumably a reference to his experiences of direct communication with God on Mt. Horeb and elsewhere.

[20] The reference here must be, not to God himself (though in Philo's Platonist models it would have been), but rather to his Logos—a concept that Philo has borrowed from Stoicism, or rather Stoicizing Platonism. That being the case, this sequence of 'negative-theological' utterances has posed problems for interpreters. I think that all Philo can mean, unless he is getting carried away by his own rhetoric, is that the Logos is superior to any individual Form, as being the whole of which they are the parts.

[21] Philo—in line here with the Stoics, and very possibly with the later Academy as well—does not, then, wish to impute to Matter, or the Receptacle, any disorderly motion of its own. For Philo, this would accord it too much of an identity. It also indicates his discomfort with the idea of a pre-cosmic chaos. In §23, admittedly, he describes it as 'unordered, devoid of quality, lacking life, dissimilar, full of inconsistency, maladjustment and disharmony'; but all this does not, I think, add up to having a disorderly motion of its own.

[22] As witnessed also by that curious piece *On the Eternity of the World*, which is best seen, I think, as one half of a *controversia*, of which Philo may or may not have ever composed the other half. The doctrinal position seems to be derived most immediately from a work of the second-century Peripatetic Critolaus. At least it shows that Philo is thoroughly abreast of the arguments on this issue.

philosophy (see also Tieleman, this volume, esp. §1), and that is surely true of Philo. Let us turn next to §§26–27, where he tries to work out an acceptable solution:

> When he says that 'in the beginning, God made the heaven and the earth', he does not take the term 'beginning' (ἀρχή), as some people think, in a temporal sense. For there was no time before the cosmos, but rather it either came into existence together with the cosmos or after it. When we consider that time is the extension of the movement of the cosmos, and that there could not be any movement earlier than the thing that moves but must necessarily be established either later or at the same time, then we must necessarily conclude that time too is either the same age as the cosmos or younger than it. To venture to affirm that it is older is unphilosophical.
>
> If 'beginning' in the present context is not taken in a temporal sense, it is likely that its use indicates beginning in the *numerical* sense (κατ' ἀριθμὸν), so that the expression 'in the beginning he made ...' is equivalent to 'he *first* made the heaven.' It is indeed reasonable that heaven should in fact be the first thing to enter into becoming, being both best of all created things and made from the purest substance, because it was to be the holiest dwelling-place for the manifest and visible gods. Even if he who made it proceeded to make all things simultaneously, it is nonetheless true that what comes into existence in a fine way (καλῶς) did possess order (τάξις), for there is nothing fine in disorder. Order is a sequence and series of things that precede and follow, if not in the completed products, then certainly in the conceptions of the builders.[23] Only in this way could they be precisely arranged, and not deviate from their path or be prone to confusion.

We see Philo here wrestling with what is for him a serious problem. He is certainly unwilling to give up the principle that the world is created by God; but his philosophical training alerts him to the problem of postulating creation at a point in time, with the attendant problem of the status of a pre-existing chaotic 'matter'. The device of postulating priority in order, or dignity, instead of time, will only work if one abandons, either explicitly or tacitly, the notion that there was a stage when the world was not, and when the Creator brought it into existence. This is something that Philo is quite unwilling to do in any explicit way, so we are left with a conundrum which has exercised the minds of the chief authorities in the field.[24]

---

[23] That is to say, even if the whole physical cosmos were created simultaneously, there would still be a τάξις of prior and posterior entities in the guiding plan of the Creator.

[24] Such as, for instance, H.A. Wolfson, *Philo: Foundation of Religious Philosophy in*

It seems to me that, if we have to choose, Philo actually comes nearest to a version of the original defence of Plato's position put forward by Speusippus and Xenocrates, which Aristotle criticizes in the *De caelo*. Philo, after all, holds (as becomes apparent a little further on in the *De opificio mundi*, §§29–30) that what Moses is referring to in Gen 1:1–5 is actually God's generation (from all eternity) of the intelligible archetypes of heaven and earth, and other elements of the cosmos, such as water, *pneuma*, and light, which are the contents of his Intellect, or Logos, and which are then (but only in a logical sequence) projected onto Matter to form the physical cosmos (see also Van Kooten, this volume, §§1.1 and 1.2). This is a process which strictly takes no time. Philo is insistent, back in §13, that God did not require six days, in a literal sense, to create the world, 'for we must think of God as doing all things simultaneously (ἅμα γὰρ πάντα δρᾶν εἰκὸς θεόν)'—so really all we are left with is a notion of logical succession, based on the relative degrees of excellence of the things created. This presents a different emphasis from that of the original Academics, whose point of comparison was rather the construction of geometrical figures, in which the component lines, say, are logically prior to the completed triangle, but certainly not superior in dignity. Probably Philo would not have rejected Crantor's sense of γέγονεν either—certainly the physical cosmos is dependent upon a higher cause, external to itself—but that is not the thrust of his argument in the *De opificio mundi*.

3. *Conclusion*

There is much more that could be said on this subject, but I hope I have indicated here both that there had been a good deal of discussion in the Hellenistic schools, in the wake of Plato's *Timaeus* and Aristotle's response in the *De caelo*, as to the logical and ontological status of the physical world, and that Philo was pretty well acquainted with the ins and outs of this. Not only his exposition in the *De opificio mundi*, but the (hostile) account of Plato's position, and of Academic defences of it, that he provides (in his *persona* as a defender of Aristotle) at the beginning

Judaism, Cambridge, MA 1947, vol. 1, 300–310; David Winston, *Philo of Alexandria: The Contemplative Life, The Giants, and Selections*, New York & Toronto 1981, 10–21; Runia, *Philo of Alexandria and the* Timaeus *of Plato*, Leiden, 1986, 287–289; Roberto Radice, *Platonismo e creazionismo in Filone di Alessandria*, Milan 1989, 247–250.

of the *De aeternitate mundi* (§§13–16), fully demonstrate this. He has to balance this, however, with his stance as a pillar of the Jewish faith and of Jewish culture generally, within an Alexandrian milieu, and this inevitably serves to obscure his position.

GOD THE CREATOR, GOD THE CREATION: NUMENIUS' INTERPRETATION OF GENESIS 1:2 (FRG. 30)

Robbert M. van den Berg

1. *Introduction*

At the head of the harbour there stands an olive tree with spreading leaves, and near it is a misty and pleasant cave sacred to the nymphs called naiads.[1]

When in Homer's *Odyssey* the hero Odysseus has finally, after many wanderings and hardships, made his way home, he cannot believe his luck. The goddess Athena has to convince him that this really is his beloved Ithaca by pointing out some of its most prominent features, including a cave dedicated to water nymphs, the so-called Naiads. The passage is a wonderful gem of Greek poetry, evoking a lively picture of what looks like a pleasant refuge from the burning Greek summer sun. Yet to ancient Platonists it also presented a profound exegetical puzzle. On the assumption that Homer was a great sage, they set out to discover the hidden wisdom embedded in these and the subsequent verses.

One such allegorical interpretation was composed by Porphyry (234 – c. 304 AD), one of the big names in the history of Neoplatonism.[2] Not only was he one of the most important students of its founder, Plotinus, whose work he edited, but he was also a major philosopher in his own right, whose work on Aristotle was to exercise a profound influence well into the Middle Ages. The pivotal idea around which his interpretation of Homer's description of the cave of the nymphs hinges is that the cave is a symbol of the cosmos (see also Van Kooten, this volume, §2.6.7) and that Homer was thus engaged in a piece of cosmology.

[1] Homer, *Odyssey* 13.102–104; transl. from Seminar Classics 609, State University of New York at Buffalo, *Porphyry: The Cave of the Nymphs in the Odyssey* (Arethusa Monographs 1), Buffalo 1969, 3.

[2] Edited with an English translation by Seminar Classics, *Porphyry: The Cave*; for the same edition with French translation, see G. Lardreau and Y. Le Lay, *Porphyre: L'antre des nymphes dans l'Odyssée*, Lagrasse 1989.

Porphyry did not develop his cosmological interpretation from scratch. As he freely acknowledges, he derives much from his predecessors, especially Cronius and Numenius, two figures counted among the so-called Middle Platonists, who differ from the Neoplatonists especially where the status of the first principle is concerned. In spite of this and other differences though, some of them were held in high esteem by the first Neoplatonists, even to the degree that some contemporaries accused Plotinus of having plagiarized Numenius, whereas about Porphyry it was remarked that it was a cause of amazement to find that he said something different from Numenius. One should keep in mind that these are polemical accusations, which do not do justice either to the genius of Plotinus nor to the intellect of Porphyry, but still it says something about their intimate relation to Numenius.[3] It is this philosopher, about whose personality we know unfortunately little more than that he was from Apamea, Syria, and (probably) active in the second century AD, who will be at the centre of interest in this contribution.[4]

About Numenius' interpretation of the nymphs, the Naiads, Porphyry reports:

> In a stricter sense it is the powers presiding over waters that we call naiad nymphs, but they (sc. the ancients) also give this name to all the souls in general descending into the world of becoming. For they thought that the souls settle by the water, which is divinely inspired, as Numenius says; in support of this he cites the words of the prophet,
>
> 'the Spirit of God was borne upon the waters (Gen 1:2).'
>
> The Egyptians as well, he says, represent daemons as standing not on solid ground but on a boat; this applies to the Sun and, in short, to every one of them. We must understand that these represent souls hovering over moisture, i.e. those souls descending to the world of becoming.
>
> It is for the same reason, Numenius says, that Heraclitus says that 'it is a delight, not death, for souls to become moist', meaning that the fall into the world of becoming is a pleasure for them; and in another place, 'We live their death, they live our death'. And he believes that this is the

[3] For the accusation of plagiarism directed against Plotinus, see Porphyry, *Vita Plotini* 17. For Porphyry, see Proclus, *In Platonis Timaeum commentaria* I.77, 22ff. (= Numenius, frg. 37) with the comments of J.H. Waszink, 'Porphyrios und Numenius', in: Fondation Hardt, *Entretiens sur l'antiquité classique*, vol. 12: *Porphyre*, Genève 1966, 33–83, esp. 35–36.

[4] On Numenius and his philosophy, see the introduction to E. des Places, *Numénius: Fragments*, Paris 1973; J. Dillon, *The Middle Platonists: A Study of Platonism 80 B.C. to A.D. 220*, London 1996[2] (revised edition with new afterword), 361–383 and 448–449; M. Frede, 'Numenius', in: W. Haase (ed.), *Aufstieg und Niedergang der römischen Welt*, vol. II.36.2, Berlin 1987, 1034–1075.

> reason why Homer calls those in the world of becoming *dieros*, because they have humid souls.[5] Blood and moist seed are dear to those souls;[6] the souls of plants, on the other hand, are nourished by water (Porphyry, *De antro nympharum* 10 = Numenius, frg. 30; transl. [adapted] Seminar Classics, *Porphyry*).

In this interpretation, 'water' represents the 'world of becoming', γένεσις in Greek. As we shall see, the Platonists oppose this to the unchanging world of Being, the domain of the Platonic Forms. As we noted above, both Porphyry and Numenius assumed that the cave of the nymphs represents our cosmos, the world of becoming. According to them, the water nymphs to whom the moist cave is dedicated are some sort of lower deities that descend from a higher realm into the world of becoming in order to preside over it. Then as now, commentators back up their interpretations by citing parallel passages and Numenius had collected at least three of them.

First of all, there are the words of the Moses, the 'prophet', who writes in Gen 1:2 that in the beginning the spirit of God hovered over the waters. Secondly, the Egyptians depict the Sun and other daemons as standing on a boat. Apparently, Numenius was acquainted with the images of Re the Egyptian sun god, abundantly present on the walls of Egyptian temples and on papyri, who together with his fellow gods by day travels through the sky on a bark, just as the Greek god Helios does on a chariot. Contrary to Numenius' interest for Jewish material, he shared his curiosity for the Egyptians with many other Platonists, such as the Middle Platonist Plutarch and the Neoplatonist Iamblichus.[7] Thirdly, Numenius refers to the cryptic Greek philosopher, Heraclitus,

---

[5] A little pun on the Greek word διερός, which may either mean 'alive' or 'wet'. Since it is the soul which makes the body alive, and since Platonists compare the matter of which the body consists to water (see §3 below), the soul that descends to live in a body is *dieros* in the two senses of the word. The reference is to Homer, *Odyssey* 6.201. Cf. Cornutus, *De natura deorum*, p. 45, 3: διερὸν γὰρ τὸ ὑγρόν ἐστιν (discussing Aphrodite and human procreation).

[6] According to the views hold by Aristotle and many others on the process of human procreation, the male sperm contains the human form which is embedded in female menstrual blood as its material substratum; the combination of the two results in the human embryo. Numenius, frg. 36 informs us that Numenius took a lively interest in these matters, as did others who 'in the manner of Pythagoras' interpreted the various rivers mentioned in such authoritative writers as Plato, Hesiod, and Pherecydes as allegorical descriptions of the animation of the human embryo.

[7] Plutarch e.g. observes that in Egyptian iconography the Sun and the Moon travel by boat, not by chariot as Helios does (Plutarch, *De Iside et Osiride* 364 C–D); cf. Iamblichus, *De mysteriis* VII.2.

whose saying, 'it is a delight, not death, for souls to become moist' is interpreted as a reference to the descent of souls into the watery realm of becoming.

This quotation of Genesis is famous for being the first attested allegorical interpretation of the Bible outside Jewish-Christian circles.[8] The goal of this paper is to elaborate on Numenius' interpretation and place it in its context. Unfortunately his work has only survived in bits and pieces as quotations, often out of context, in later Neoplatonic and Patristic writings,[9] yet the surviving material throws an interesting light on this passage.

2. *The background of Numenius' allegorical interpretation*

Let us first reflect a bit on the background of Numenius' allegorical interpretation. Why when commenting on Homer did he think it necessary to quote Jewish scriptures in the first place?

Numenius' (assumed) profound knowledge of Judaism, his apparent sympathy towards it in combination with the fact that Apamea is known to have been home to a flourishing Jewish community, has led some scholars, especially of older generations, to speculate whether in fact Numenius himself was a Jew. As such, the scanty evidence does not allow us to exclude this possibility, yet neither does it in any way confirm this hypothesis, as more recent scholarship stresses. One article which scrutinizes the evidence for Numenius' acquaintance with Judaism takes this position to its logical extreme by denying Numenius even anything more than only 'a fragmentary and unreliable knowledge of Jewish religion' of the sort that in the second century 'was disbursed among men to whom the ancient literature of Israel was as strange as its unknown God'.[10]

---

[8] So e.g. E.R. Dodds, *Pagan and Christian in an Age of Anxiety: Some Aspects of Religious Experience from Marcus Aurelius to Constantine*, Cambridge 1965, 130 note 2.

[9] The standard edition of these fragments is Des Places, *Numénius: Fragments*.

[10] M.J. Edwards, 'Atticizing Moses? Numenius, the Fathers and the Jews', *Vigiliae Christianae* 44 (1990) 64–75, esp. 73, who gives a most useful discussion of all (supposed) Jewish elements in the fragments of Numenius. For examples of the opposite view in older literature, see e.g. H.Ch. Puech, 'Numénius d'Apmée et les théologies orientales au second siècle', in: J. Bidez, *Mélanges Bidez*, vol. 2 (Annuaire de l'Institut de philologie et d'histoire orientales), Brussels 1934, 745–778, esp. 754: 'Une question brûle alors les lèvres: cet admirateur du judaïsme ne serait-il pas lui-même un juif? … S'il n'est donc pas certain que notre Numénius soit juif de naissance, du moins pencherais-je à en faire

For a correct assessment of Numenius' relation towards Judaism, it is important to understand the background of Numenius' interest in Jewish Scriptures in relation to what philosophy is about. A remark that comes from the first book of his *On the Good* is programmatic:

> Now that we have mentioned Plato's testimonies concerning this issue (sc. that of God) and examined it on the basis of these, we need to move backwards and connect these to the words of Pythagoras, and call upon the peoples of renown, adding their rituals, their doctrines and their foundations which they established in accordance with Plato, such as those established by the Brahmans, the Jews, the Magians, and the Egyptians (Numenius, frg. 1a).

Numenius here appears to do philosophy in a way characteristic for his time.[11] Rather than thinking a problem through for himself, he puts his trust in authoritative thinkers from the past, especially in Plato and ultimately in Pythagoras, since he supposed that the philosophy of Plato went back on that of Pythagoras. It is for this reason that Numenius was commonly referred to in Antiquity as a Pythagorean, even though most of his philosophy deals with Plato's writings. But Numenius and his contemporaries did not just consider philosophy to be the study of ancient *Greek* wisdom. It was commonly assumed that the Truth had been available to all, Greeks and Barbarians alike.

What is more, it was also believed that primitive man had had a better access to the Truth, since in primitive times mankind had not yet been corrupted and that therefore the older the knowledge, the better it was. For this reason, peoples which were supposed the be older than Greeks, such as the Egyptians, were hold to be especially wise. Unfortunately, though, this primitive knowledge was not directly available. It had been passed down the generations under the guise of myths and rituals and could only be uncovered by means of allegorical interpre-

un sémite'; E. des Places, 'Numénius et la Bible', in: E. des Places, *Études Platoniciennes 1929–1979*, Leiden 1981, 310–315 (originally published in: *Homenaje a Juan Prado*, 1975), who concludes (p. 315): 'Connaissance du milieu juif, sympathie pour les Ecritures des Hébreux, voilà qui met Numénius à part dans le platonisme moyen ou, si l'on veut, le "prénéoplatonisme" dont il est le principal représentant.'

[11] This characteristic of post-Hellenistic philosophy has most recently been discussed by G.R. Boys-Stones, *Post-Hellenistic Philosophy: A Study of its Development from the Stoics to Origen*, Oxford 2001 (for Numenius, frg. 1a, see p. 116), an interesting yet in my opinion biased book (for which see my review in *Hermathena* 170 [2001] 104–107), yet it has been recognized much earlier, see e.g. the discussion of Numenius' respect for various ancient authorities by Waszink, 'Porphyrios', 45–48, Puech, 'Numénius', and of course the vivid picture of the second century sketched by A.-J. Festugière in the introduction to his monumental *La révélation d'Hermès Trismégiste*, vol. 1, Paris 1944, chaps. one and two.

tation. Allegory is a slippery slope, notorious for yielding endless series of different interpretations, so some sort of criterion for the correctness of an interpretation had to be set up. This became the harmony of the various traditions: if the allegorical interpretation of, say Homer's cave of the nymphs, could be shown to be in harmony with the myths of Jews and Egyptians, it had probably to be right.

In the light of speculations about Numenius' Jewishness, it is important to note that Numenius starts his investigations from Plato, who apparently functions as his point of reference.[12] The correctness of Plato's views is demonstrated by the fact that he sings the same song as peoples more ancient than the Greeks, including the Jews and their authoritative sage Moses. Whether or not he had a Jewish background —perhaps tellingly Numenius always refers to the Jews in the third person—he presents himself here first and foremost as a student of Greek philosophy, of Plato and Pythagoras, thus as someone partaking in the Greek tradition. We may contrast this to the Jew Philo of Alexandria, who in his allegorical interpretations of the *Pentateuch* also connects Platonic philosophy to Jewish wisdom, but this time from the Jewish perspective.

Yet, did Numenius perhaps assign a special position to Jewish wisdom? After all, Numenius is famous for having said:

> What is Plato, but Moses speaking in Attic Greek? (Numenius, frg. 8.13)

Why did he not say that Plato was a Brahman, Magian, or Egyptian speaking in Attic Greek? The truth of the matter is that we do not know whether he said something of the sort. This quotation comes from the Church Fathers, who quote Numenius for their own apologetic purposes. They reasoned that if, as the Greeks themselves admitted, there is a correspondence between Greek and Biblical wisdom, and, that if the older your source, the better, then, given that Moses predates Plato and the lot, it follows that Christianity, which has its roots in the wisdom of Moses, is superior to Hellenic culture. In support of this line of argument, the Fathers quote from Greek philosophers, such as Numenius, in order to show that the latter consent to the premises of their argument, and hence are forced to agree with the conclusion. Whatever Numenius thought about Plato's relations to Brahmans and other peoples did not interest them, neither did the precise intention of Numenius' words.

---

[12] As Waszink, 'Porphyrios', 46 observes.

In short, I think that Numenius was a Greek philosopher with an interest typical of his age in other sources of wisdom. The surviving material does not allow us to say anything about the depth or superficiality of his knowledge of these sources. At any rate, the fact that he knows Gen 1:2 proves little, since other contemporary Greek writers such as Galen and Pseudo-Longinus refer to it as well.[13]

One final word about the relation between Philo of Alexandria and Numenius. Since Numenius is on record for being responsible for the introduction of the allegorical interpretation of the Old Testament to the Hellenic world, it is sometimes suggested that he drew on Philo of Alexandria. With reference to Numenius' interpretation of Gen 1:2, the passage that concerns us here, it has been pointed out that like Philo he interprets a book which is not Greek in the light of Greek philosophy.[14] Perhaps the possibility cannot be excluded, but there is little that positively supports this assumption. As has been observed, the problem of the possible dependence of the exegetical thought of Numenius on Philo is in general a complex one. One the one hand, we know little of the study of Philo outside Jewish and Christian communities. On the other, as we have seen above, allegorical exegesis was commonly practiced at the time of Philo and Numenius, and eventual similarities may well be explained by the use of this common approach.[15] In fact, as some contributions to this volume show, even the earliest Christians connected biblical narratives to contemporary philosophical thought, apparently independently from Philo.[16] To these general considerations, I would like to add that, if anything, Numenius' interpretation of Gen 1:2 rather suggests that Numenius was *not* depended on Philo. Philo gives two different interpretations of Gen 1:2 According to the one, God is engaged in the act of creating the intelligible world, that of the Platonic Ideas. According to an alternative interpretation, this spirit is the third element, air.[17] Obviously, then, neither of these resembles what we find in Numenius.

---

[13] Pseudo-Longinus, *On the Sublime* 9.9. For Galen, see the contribution of T. Tieleman to this volume.

[14] Dodds, *Pagan and Christian*, 130 note 2; Waszink, 'Porphyrios', 50–51.

[15] As R. Lamberton, *Homer the Theologian: Neoplatonist Allegorical Reading and the Growth of the Epic Tradition*, Berkeley 1986, 75 rightly remarks.

[16] See e.g. the contributions by E. Adams and G.H. van Kooten to this volume.

[17] For the former interpretation, see Philo, *De opificio mundi* 30 discussed by J. Dillon in his contribution to this volume (cf. also the commentary by D.T. Runia, *Philo of Alexandria: On the Creation of the Cosmos according to Moses; Introduction, Translation and Commentary*, Leiden 2001, 166–167); for the alternative interpretation see e.g. *De gigantibus* 22.

3. *Numenius' interpretation against the background of his philosophy*

In this section I shall try to add some flesh to the bare bones of Numenius' interpretation of Gen 1:2 by placing it against the background of his philosophy. Numenius is best known for his theory of the three gods to whom he also refers as Intellects. The first god, whom he calls Father (Πατήρ), is the Good or Being itself. It is the source of all Being, i.e. the unchanging world of the Platonic Forms and is the ultimate source of all beauty. The second God is the Creator (Ποιητής) or Demiurge (Δημιουργός, literally the 'Craftsman'), the one who creates the world of Becoming, this ever-changing cosmos of ours, after the image of the world of Forms. This cosmos, finally, is the third god and is called the Creation (Ποίημα).[18]

As was said in the introduction, Numenius does not pretend to develop any new insights by himself. Whatever he says can be traced back to such authorities like Plato, or so he wants us to believe. One source of inspiration for this doctrine of the three Gods is the so-called *Second Letter* by Plato.[19] Modern scholars almost unanimously assume that this letter is a fraud, but in Antiquity its authenticity was beyond doubt. It was even considered as a key text for the understanding of Plato. An enigmatic passage from it reads:

> Around the king of all things all things are and they are because of him and that is the cause of all beautiful things. Around the second are the secondary things, and around the third the tertiary things (Ps.-Plato, *Epistulae* II, 312E1–4).

Below, we shall return to this text when we come to discuss Justin Martyr's interpretation of Gen 1:2. Numenius' other main source of Platonic inspiration is Plato's *Timaeus* in which Plato presents us with a story how a divine craftsman, the Demiurge creates our cosmos after the example of the world of Platonic Forms. In this account Plato calls our cosmos a perceptible god (*Timaeus* 34B; 92C), since it was created in the image of the divine Forms by a divine Craftsman, whom Numenius clearly identifies with his second god.

As such, this scheme of three Gods is clear enough, yet Numenius complicates things by claiming that the second and third god, the Cre-

---

[18] See e.g. Numenius, frgs. 16 and 21.

[19] Ps.-Plato, *Epistulae* II, 314C; cf. Numenius, frg. 24.51ff. and Des Places, *Numénius: Fragments*, 10.

ator and the Creation are actually one, or perhaps better *is* one.[20] Whereas the first god remains unchanged, turned towards himself at his own level, the second god is torn apart as a result of his creative activities. The second god does not create *ex nihilo* (see also Tieleman, this volume, §1), but uses pre-existing matter in which he projects the Platonic Forms, thus creating an image of the world of Forms.[21] However, as most Platonists before and after him, Numenius considers matter as the source of all evil. Whereas the first god is just turned towards himself, the second god has to look into two opposite directions. On the one hand, he looks upwards to the Forms, which serve him as a model for his creating activity. Meanwhile, since he embodies these Forms in matter, he has to look downwards to matter. It is here that tragedy strikes: while caring for his creation, the second god is seduced by matter. He forgets about himself and about his transcendent status, since he identifies himself with his material creation. In short, he is torn apart and becomes both the divine Creator and the divine Creation.[22] Since matter as such is evil, the creation is divine to the extent that God is present in it and since according to Plato in the *Timaeus*, it is the World Soul which makes the cosmos a divine being, it has plausibly been suggested that Numenius' third God is the Platonic World Soul.[23]

Let us now return to Numenius' interpretation of the cave of the nymphs. The cave, our universe, is a damp place, since water is a standard symbol of matter in the Platonic tradition. In the existing fragments Numenius returns to this symbol time and again. Matter is compared to a rapidly streaming river (frg. 3) to bring out that the material world lacks form and stability, as opposed to the eternal changeless Platonic Forms. This is the world of Heraclitus' *panta rhei*. In the same vein, Numenius (frg. 52.33–35) qualifies the formless ever-changing matter that the Demiurge in Plato's *Timaeus* finds when he sits down to create our world as 'fluid': nothing was 'at rest, but in discordant and unordered motion' (*Timaeus* 30A; cf. 52D–53B). The

20 Frede, 'Numenius', 1057: 'der Gott aber, der zweiter und dritter Gott ist, ist einer'. This remark is meant as a criticism of Dillon, *The Middle Platonists*, 367, yet, it mainly reinforces Dillon's interpretation (cf. Dillon, *The Middle Platonists*, 448).

21 In fact an unambiguous doctrine of *creatio ex nihilo* was absent even from the minds of contemporary Christians. On this issue, see further D.T. Runia, 'Plato's *Timaeus*, First Principle(s), and Creation in Philo and Early Christian Thought', in: G.J. Reydams-Schils (ed.), *Plato's Timaeus as Cultural Icon*, Notre Dame, Indiana 2003, 133–151.

22 See e.g. Numenius, frg. 11.

23 For this identification, see e.g. Dillon, *The Middle Platonists*, 374; Frede, 'Numenius', 1068–1069.

comparison with God the Creator hovering over the chaos in Gen 1:2 commends itself almost automatically.

The Jewish God and the Egyptian daemons do not really get their feet wet. The former hovers above it, the latter stand on boats. Numenius probably wanted his reader to take notice of this. In one of the fragments he compares the Demiurge to a helmsman who is explicitly said to be above the matter 'as if above a ship upon sea':

> A helmsman, after all, sailing on the high seas, seated high above the tiller, directs the ship from his perch, and his eyes and his intellect are straining upwards to the aether, towards the heights of heaven, and his route comes down to him from above through the heavens, while he sails below on the sea; even so the Demiurge, binding matter fast by harmony, so that it may not break loose or wonder astray, himself takes his seat above it, as if above a ship upon the sea; and he directs the harmony, steering it with the Forms, and takes his critical faculty from this contemplation, while he derives his impulsive faculty from his desire (Numenius, frg. 18; transl. Dillon, *The Middle Platonists*, 370).

This comparison brings out nicely the fact that the Demiurge looks both upwards and downwards, upwards to the Forms for direction like a helmsman navigates by the sun and stars; downwards to the Creation in order to care for it, 'binding matter fast by harmony' so that it does not run wild, a reference to Plato's *Timaeus* where the Demiurge is said to bind the cosmos together with unbreakable ties (31B–32C). Even though the helmsman is situated high on his ship, he is not above in the heavens, rather he is in between heaven and earth, between the first God and the third. Finally, there is the Demiurge's desire. This is the desire for matter which makes him descend to matter and which brings about the division into the second and the third god.[24] This recalls Numenius' interpretation of Heraclitus' saying that 'it is a delight, not death, for souls to become moist' (Numenius, frg. 30, see §1 above). Whereas Heraclitus may well have meant that drunkenness is a source of pleasure for souls, Numenius thinks that it refers to the descent of souls into matter.

It seems likely then that Numenius equates the god of Gen 1:2 with his second God, not with his first god. Note in this context that Numenius refers to the Egyptian counterparts of the Jewish god as daemons, i.e. as lower divine beings (though, contrary to our modern conception of demons, not necessarily evil ones). This pops the question

[24] Frg. 11.15–20.

whether Numenius believed that the Jews also venerated his first God, or not. The second opportunity can certainly not be excluded. As Numenius himself says, commenting on Plato, *Timaeus* 28C where Plato stresses how hard it is to find the maker and father of this cosmos, let alone talk about him:

> Since Plato knows that among men only the Craftsman is known, while the first Intellect, who is called Being itself, is completely unknown among them, for that reason he says this, as if someone would speak like this: 'O men, the Intellect which you focus on is not the first Intellect, but there is another Intellect before it, superior and more divine' (Numenius, frg. 17).

In fact, later Neoplatonic authors like Porphyry, the emperor Julian, and Proclus, take the Jews and the Christians to task for failing to see that their god is not the highest god (for the reciprocity of this polemic, see Van Kooten, this volume, §2.6.7 note 68). Some scholars believe to have discovered passages that indicate that Numenius thought that the Jews venerated the highest god, true Being itself, but one of them is based on a conjecture that carries little conviction,[25] whereas the other (frg. 56) only states that the Jewish god considers himself superior to all other gods and does not allow them to be worshipped.

One last question concerning Numenius' theory of creation and its interpretation of the Genesis story. According to Genesis, God creates the world order that before that moment had not existed. What can we say about Numenius' position about this? Plato in the *Timaeus* seems to suggest that the world had a beginning in time. We know that some Middle Platonists like Plutarch of Chaeroneia took this literally, yet other Platonists did not, since a beginning in time of the world left them exposed to devastating criticism. One fragment (frg. 12) seems to suggest that according to Numenius the cosmos goes through cycles of becoming and destruction. According to the myth in Plato's *Statesman*, during one period God governs the cosmos, whereas at another time he leaves the cosmos to fend for itself and returns into his conning-tower, which inaugurates a period of great upheaval, until God returns to the tiller and restores order. If so, Numenius would probably interpret the creation accounts of the *Timaeus* and Genesis as describing the

[25] Frg. 13 on one reading, followed by Des Places, identifies Numenius' first god with 'He-who-is' (Exodus 3:14), however the problematic syntax suggest that there is a textual error here; on this see e.g. Dillon, *The Middle Platonists*, 449; Edwards, 'Atticizing Moses?', 66–67.

beginning of such a cycle. However, other interpretations are possible and, as John Dillon observes, it is not clear how far frg. 12 commits Numenius to cosmic cycles.[26]

4. *A comparison to a contemporary Christian author: Justin Martyr*

As was noted in the context of the discussion of a possible link between Philo and Numenius, allegorical interpretation was a common practice at the time, as may be illustrated by the interpretation of Gen 1:2 by a Christian contemporary of Numenius, Justin Martyr (c. 100 – c. 165), which bears great resemblance to that of the Platonist.

First, by way of introduction a few words about Justin himself. According to his own, no doubt somewhat embellished account of his conversion to Christianity in the *Dialogue with Trypho* 2–5, Justin had gone through a serious education in philosophy, studying under a Stoic, a Peripatetic, a Pythagorean and a Platonic teacher respectively. Especially Platonism impressed him, and if it had not been for an encounter with an old man at the beach who showed him the shortcomings of Platonism and the superiority of Christian philosophy, he might well have ended up as a second Numenius. His attitude towards Greek philosophy is interesting, if not unproblematic. Henry Chadwick describes him as 'the most optimistic about the harmony of Christianity and Greek philosophy. For him the gospel and the best elements in Plato and the Stoics are almost identical ways of apprehending the same truth.'[27] It is optimism with a pessimistic flavour. On the one hand in *the Dialogue with Trypho*, Justin presents Platonism as the crowning achievement of Greek philosophy, not because he sees Platonism as some kind of preparation for Christianity, but, on the contrary, because it is the only serious rival to true philosophy, i.e. Christianity (cf. also Van Kooten, this volume, §§2.6.3 and 2.6.4 note 55). It is only after the old man has destroyed Platonism that Justin becomes receptive to Christianity. Justin's reason for rejecting Greek philosophy is interesting: the Jewish prophets are men of far greater antiquity than any of the Greek philosophers (*Dialogue* 7.1). So, just like Numenius, Justin believes that antiquity implies validity.[28]

---

[26] Dillon, *The Middle Platonists*, 370.

[27] H. Chadwick, *Early Christian Thought and the Classical Tradition: Studies in Justin, Clement, and Origin*, Oxford 1984 (1966[1]), 10–11.

[28] I thank W. Otten for drawing my attention to O. Skarsaune, 'The Conversion of Justin Martyr', *Studia Theologica* 30 (1976) 53–73.

On the other hand, in his *Apologies*, where Justin sets out to show that Christianity is not the repugnantly exotic movement that some of its detractors claim it to be, he takes a somewhat different line. Greek philosophy and the Jewish prophets often appear to say the same things. Justin explains this by assuming that in some of those cases the Greek philosophers had correctly used their divinely given reason, and that in other cases they had taken a leaf from the books of the Jewish prophets, even though they might at times have failed to understand completely what they were copying. In short, even though, Justin adopts the ideas about an ancient tradition of true philosophy, which we have found at work in Numenius for his own ends.[29]

One of those instances where Greek philosophy is in harmony with the Jewish prophets is the case of Plato's account of the origin of the world as he describes it in the *Timaeus*. In his *Apologies* 1.59–60, Justin sets out to show that Plato derived his account of the creation of the world from 'the first of the prophets, Moses, … and older than the Greek writers through whom the prophetic Spirit has revealed how and from what materials God created (ἐδημιούργησεν) the world in the beginning through his word (λόγος) out of the chaotic earth and waters' (*Apologies* 1.59). Note that Justin stresses that Moses is older than the Greek writers, and hence is superior.

Turning to Plato's account in the *Timaeus*, Justin recognizes in it a Christian trinity (*Apologies* 1.60).[30] Apparently Plato's Demiurge is equated with the God from Genesis, the first member of the Christian Trinity. This reminds us of the criticism voiced by later Neoplatonists that the Christians mistake the Demiurge for the highest God. The Son, the second member of the Trinity, is equated with the entity which according to Plato, *Timaeus* 36B envelops the cosmos in the manner of the Greek letter *chi* (X). According to Justin this *chi* is nothing else than the Christian cross misunderstood and misconstrued (cf. also Otten, this volume, §5a). Plato's biblical source for it is the story of how Moses saves the Israelites from venomous snakes by making a cross from bronze, an allusion to Numbers 21. Apparently, Justin is so anxious to see parallels between Plato and Moses that he allows

[29] For Justin on pagan philosophy and its relation to true Christian philosophy, see Boys-Stones, *Post-Hellenistic Philosophy*, 184–188.

[30] On this passage, see H.D. Saffrey and L.G. Westerink, *Proclus: Théologie Platonicienne*, vol. 2.2, Paris 1974, XXXIX–XLI, who call it a form of 'concordisme le plus détestable et le moins convaincant' (p. XL).

himself some liberty where the actual texts are concerned. According to Numbers 21, Moses orders to make a serpent, not a cross[31] and Justin's quotation of Plato's *Timaeus* lacks from the text, which we have.[32] What is more, he does not understand or wish to understand the nature of this *chi*-shaped entity. Plato clearly has the World Soul in mind here. For Justin however, it is something different from the entity that ensouls the cosmos which he equates with the third member of the Trinity.

About this third member Justin has the following to say:

> The third place he (sc. Plato) assigns to the 'Spirit which was borne upon the waters', when he says that 'the tertiary things are around the third'.

Justin here quotes from Pseudo-Plato's *Second Letter*, the text which, as we have seen (see §3 above), was so important in Numenius' theology of the three Gods. Numenius' third God, it will be remembered, is the World Soul, which Justin probably associated with the Spirit of God.[33] This text shows that Numenius' interpretation of Genesis against the background of Platonic philosophy was not unique in his day.

Did Justin arrive at his interpretation independently, or is there some link between Numenius and Justin? We have after all seen that Numenius was a highly influential philosopher and that Justin had enthusiastically studied Platonism before his conversion to Christianity.[34] Fact is that in the wake of Justin, Pseudo-Plato's *Second Letter* remained a favourite text among those Christian writers who wished to stress the resemblance between Plato and Christianity.[35] However, Platonism is not Christianity and Justin was quite right to have some second thoughts about the harmony between the two. It is an article of Platonic faith that this cosmos, or at least its soul, is a god, a doctrine

---

[31] As G.H. van Kooten kindly points out to me, Justin probably has John 3:14–15 in mind here: 'Just as Moses lifted up the serpent in the wilderness, so the Son of Man must be lifted up (i.e. crucified), in order that everyone who has faith may in him have eternal life' (transl. Revised English Bible).

[32] Justin, *Apologies* 1.60.1: (Plato) λέγει· Ἐχίασεν αὐτὸν ἐν τῷ παντί.

[33] On the interpretation of the 'World Soul' in ancient philosophy and Christianity, see H. Ziebritzki, *Heiliger Geist und Weltseele: das Problem der dritten Hypostase bei Origenes, Plotin und ihren Vorläufern*, Tübingen 1994.

[34] That Justin was influenced by Numenius is e.g. suggested by L. Brisson, 'The Platonic Background in the *Apocalypse of Zostrianos*: Numenius and *Letter II* attributed to Plato', in: J.J. Cleary (ed.), *Traditions of Platonism: Essays in Honour of John Dillon*, Aldershot 1999, 173–188.

[35] This reception has been studied in great detail by Saffrey & Westerink, *Proclus*, XX–LIX ('Histoire des exégèses de la *Lettre II* de Platon dans la tradition platonicienne').

utterly unacceptable to Christians: already the author of the Pauline *Letter to the Colossians* 2:8 had cautioned his fellow Christians not to fall prey to (a deceptive kind of) philosophy and the veneration of the cosmos. Warnings of this type were repeated time and again, including by such influential figures as Augustine, although the Platonic spirit was never fully driven out of Genesis.[36]

[36] On the topic of cosmic worship in ancient Platonism and the rejection of it by Christianity, see J. Pépin, 'Cosmic Piety', in: A.H. Armstrong (ed.), *Classical Mediterranean Spirituality: Egyptian, Greek, Roman*, London 1986: 408–435; on Augustine see esp. p. 432 note 3, where Pépin cites Augustine's rejection of the interpretation of the 'Spirit of God borne upon the waters' as the World Soul (in *De diversis quaestionibus ad Simplicianum* 2.1–5). On Graeco-Roman cosmology in the Pauline writings, see also as G.H. van Kooten, *Cosmic Christology in Paul and the Pauline School: Colossians and Ephesians in the Context of Graeco-Roman Cosmology*, Tübingen 2003, esp. 65–66 and 145–146 on the veneration of the cosmos. On Philo's criticism of those who 'show more reverence for the cosmos than is its due', see also Dillon, this volume, §2.

GALEN AND GENESIS

Teun Tieleman

1. *Introduction*

In 1949 the distinguished scholar Richard Walzer published a monograph in which he discussed six testimonies from Galen of Pergamon (129–c. 213 CE) concerning Jewish-Christian beliefs and practices.[1] Three of these intriguing passages had only recently been spotted by Western scholars in medieval Arabic versions of Galenic treatises.[2] The others came from treatises preserved in the original Greek that had long been accessible to readers in the Christian world.[3] From these assembled pieces of evidence Galen emerges as a reasonably unbiased observer, who criticizes the Jews and Christians for having faith in undemonstrated laws and miracles but also commends the Christians for their virtuous conduct. He seems to be well informed. The second and by far longest of Walzer's texts, from the *On the Usefulness of Parts* XI.14, refers to the first chapters of Genesis. Other testimonies indicate that Galen engaged in oral discussion with Christians in Rome.[4] Galen was clearly interested in their creed and must have learned from these conversations also.

Galen's evidence is not extensive but it is precious in view of its date and content. Of particular interest is the passage from the *On the Usefulness of Parts*, which considers at some length a doctrinal issue, viz.

---

[1] R. Walzer, *Galen on Jews and Christians*, London 1949, 101 pp.

[2] A passage from Galen's *On Hippocrates' Anatomy* cited by Ibn al-Matran, *Life of Galen*, vol. I, p. 77, ed. Müller (~ Reference No. 1 Walzer), another one from his *On the First Unmoved Mover*, also cited by Ibn al-Matran, *ibidem* (Ref. No. 5 Walzer), and one from the *Summary of Plato's Republic*, as quoted by Abu'l-Fida, *Universal Chronicle*, p. 108, ed. Fleischer (~ Ref. No. 6 Walzer). For the (translated) text see the Appendix below.

[3] These are: *On the Usefulness of Parts* XI.14 (vol. 2, pp. 158–160, ed. Helmreich ~ Reference No. 2 Walzer) and *On the Differences between Pulses* III.3, II.4 (VIII, pp. 657, 579, ed. Kühn ~ Ref. Nos. 3, 4 Walzer). For the (translated) text see below, p. 132 and the Appendix.

[4] See esp. the passages from *On the Differences between Pulses* (see previous note), i.e. Walzer's Reference Nos. 3 and 4, and the passage from Eusebius, *Historia ecclesiastica* 28.13–14, quoted below, p. 143.

creation. In this paper I want to focus on this passage, taking issue with Walzer's reading according to which Galen refers to the Judaeo-Christian doctrine that God created the universe out of nothing, i.e. the doctrine of *creatio ex nihilo*. This reading has been accepted by Gerhard May in his full-scale treatment of the origins of this doctrine.[5]

That God created the world out of nothing seems the most natural way of reading the opening chapter of Genesis. However, as May himself rightly stresses,[6] we must exercise caution when we come across the statement that God created the world out of nothing. Early sources in which this statement is found may merely express the idea of God's omnipotence. In such cases *creatio ex nihilo* in its technical sense is not in play. This is generally believed to have resulted from the debate between pagans and Christians in the second century CE—which makes Galen an important witness.[7] Indeed, it seems to have been designed in conscious opposition to a fundamental assumption of the Greek philosophical tradition (cf. also Dillon, this volume, §2). From Parmenides (fifth century BCE) onward it had been axiomatic for Greek philosophers that nothing comes into being from not-being.[8] Accordingly, Plato in his extremely influential *Timaeus* pictures the divine Craftsman ('Demiurge') as bringing order to a *pre-existing* entity called the 'Receptacle' or 'Mother of Becoming' or 'the Place', which was soon identified by Plato's readers with Aristotle's material cause (see below, p. 133). This entity prevents God's best intentions from being completely realized, thereby explaining such imperfections as remain

---

[5] See G. May, *Schöpfung aus dem Nichts: Die Entstehung der Lehre von der* Creatio ex Nihilo, Berlin & New York 1978, 49 note 37, 158 note 41, 171. But cf. R.M. Grant, *Miracle and Natural Law in Graeco-Roman and Early Christian Thought*, Amsterdam 1952, 130 for a different interpretation; see further below, p. 136 note 37.

[6] May, *Schöpfung*, VIII, 8ff.

[7] The view that the doctrine originated in Hellenistic Judaism has now been generally abandoned, see H. Schwabl, 'Weltschöpfung', in: *Real-Encyclopädie der classischen Altertumswissenschaft*, Suppl. IX (1962), 1573ff.; H.-F. Weiß, *Untersuchungen zur Kosmologie des hellenistischen und palästinischen Judentums*, Berlin 1966, 146ff.; and G. Schmuttermayr, '"Schöpfung aus dem Nichts" in 2. *Makk.* 7,28?', *Biblische Zeitschrift* N.F. 17 (1973) 203–228 (cf., however, Noort, this volume, §2). Nor can Philo of Alexandria (c. 20 BCE–45 CE), who was saturated in Greek philosophy, be credited with the technical concept of *creatio ex nihilo*. Indeed, he follows a Platonic schema in positing a pre-existing matter, although it resembles the Stoic material principle in being entirely passive, see e.g. *On the Creation of the Cosmos according to Moses* 8–9, 21–22 with D.T. Runia, *Philo of Alexandria: On the Creation of the Cosmos according to Moses; Introduction, Translation and Commentary*, Leiden 2001, *ad loc.* On Philo see further May, *Schöpfung*, 9ff. and Dillon, in this volume.

[8] Aristotle (384–322 BCE) says this was the common opinion of the natural philosophers: *Physica* Α 4.187a27; *Metaphysica* Λ 7.1071b19 and 10.1075b24.

in a cosmos marked by overall purposefulness and beauty. From the Judaeo-Christian point of view, however, the postulate of the Receptacle goes against divine omnipotence. The doctrine of *creatio ex nihilo*, then, seems to be intended as the exact counterpart of the Platonic and other Greek accounts of creation that were based on the axiom that being cannot come from not-being. It is however a moot point whether it is this opposition which is at issue in the passage from the *On the Usefulness of Parts*, or in other words, whether Galen is here concerned with the concept of *creatio ex nihilo*.

Walzer's learned discussion of the Galenic material was excellent in its time and still is worth reading. Meanwhile there have been advances in research that are directly relevant to the questions at stake here. The past two decades have seen a powerful upsurge of interest in Galen and his intellectual background. We are beginning to understand Galen's double role as a medical scientist and philosopher, his individualistic type of philosophical eclecticism as well as his ideal of a unitary science of medicine-cum-philosophy.[9] More work has been done on the philosophy of the Imperial period as a whole—still badly neglected in Walzer's day. In particular we are in a better position to appreciate what Galen learned in the 'Middle Platonist' schools of his time.[10] Walzer for his part argued that important features of Galen's response to Genesis were due to the influence of the deviant Stoic Posidonius of Apamea (c. 135–51 BCE), whom he takes to have criticized his fellow-Stoics along the same lines. But since the publication of his study Posidonius' place in the history of ancient philosophy has also undergone a drastic revision.

These developments are all the more important if one considers the defective nature of our evidence. Walzer has aptly compared the

---

[9] See M. Vegetti, 'Tradizione e verità: Forme della storiografia filosofico-scientifica nel De Placitis di Galeno', in: G. Cambiano (ed.), *Storiografia e dossografia nella filosofia antica*, Torino 1986, 227–243 (see also the slightly updated English transl. in Ph. J. Van der Eijk [ed.], *Ancient Histories of Medicine: Essays in Medical Doxography and Historiography in Classical Antiquity*, Leiden 1999, 333–357).

[10] See the survey of Middle Platonism by J.M. Dillon, *The Middle Platonists*, London 1977 (repr. 1996); cf. also J. Whitaker, 'Platonic Philosophy in the Early Centuries of the Empire', in: W. Haase (ed.), *Aufstieg und Niedergang der römischen Welt*, vol. II.36.1, Berlin 1987, 81–123. On Galen and Platonism see Ph. De Lacy, 'Galen's Platonism', *American Journal of Philology* 69 (1948) 27–39; P.L. Donini, 'Motivi filosofici in Galeno', *La parola del passato* 35 (1980) 333–370; and T. Tieleman, *Galen and Chrysippus on the Soul: Argument and Refutation in the De Placitis Books II–III*, Leiden 1996, xxviiiff.

passages at issue with an actor's asides.[11] Galen provides no sustained treatment but rather takes the opportunity to insert a few comments on a topical issue. This increases the liability of our reading being conditioned by preconceived notions.

In what follows I shall first approach the central passage from its context in the *On the Usefulness of Parts* (§2). In the middle part of this paper I shall offer an analysis of the passage itself (§3). Next I shall consider other (supposed) witnesses to the debate in which Galen took part, Platonic and Stoic (§4). In the epilogue I shall put my findings into historical perspective (§5).

2. *Dissection as an act of piety: Galen's 'On the Usefulness of Parts'*

In his great work *On the Usefulness of Parts*—written in Rome between 169 and 176 CE—Galen sets out to demonstrate that each and every part of the human body has been wonderfully designed to fulfil a particular function. His approach is encapsulated in his oft-repeated adage 'Nature does nothing in vain'. It is derived from Aristotle but Galen's teleology is even more radical and in this respect closer to Stoicism.[12] In taking his teleological approach even to the humblest parts of our physique, Galen is certainly indebted to Aristotle's *On the Parts of Animals*, one of the classical treatises most dear to his heart.[13] It is pertinent to observe that he intended the work primarily for Peripatetic acquaintances, who were particularly appreciative of the Aristotelian heritage involved.[14] Galen provided an account that was scientifically more up-to-date and deploys far more anatomical knowledge than its

[11] Walzer, *Galen on Jews and Christians*, 2.

[12] On Galen's teleology see R.J. Hankinson, 'Galen explains the elephant', in: M. Matthen and B. Linsky (eds), *Philosophy and Biology* (Canadian Journal of Philosophy, Suppl. Vol. 14), Calgary 1988, 135–157; and idem, 'Galen and the best of all possible worlds', *Classical Quarterly* 39 (1989) 206–227.

[13] See P. Moraux, 'Galen and Aristotle's De partibus animalium', in: A. Gotthelf (ed.), *Aristotle on Nature and Living Things: Philosophical and Historical Studies Presented to David Balme*, Pittsburgh & Bristol 1985, 327–344.

[14] The first book was written at the request of a Peripatetic, see Galen, *On My Own Books*, chap. 1, in: I. von Müller (ed.), *Claudius Galenus: Scripta Minora*, vol. 2, Leipzig 1891, p. 96.19ff. Galen moreover tells us that the work as a whole enjoyed considerable vogue among Aristotelians, see *ibidem*, chap. 2, p. 100.18–23; cf. also T. Tieleman, 'Plotinus on the Seat of the Soul: Reverberations of Galen and Alexander in *Enn.* IV, 3 [27], 23', *Phronesis* 43 (1998) 309–310.

Aristotelian model. In so doing he built on the work of post-Aristotelian medical scientists such as Herophilus and Erasistratus (first half of the third century BCE), who were in fact the first anatomists worthy of that name. Still another important model is involved—the Platonic *Timaeus*. Like Plato, Galen presents a divine Demiurge ('Craftsman') who constructs the cosmos to the best of his ability.[15]

Galen's *On the Usefulness of Parts* can be read as a particularly extensive version of what today is called the argument from design (see also Van Woudenberg, this volume): the structure of our body provides evidence for the existence of God. As Galen memorably states, true piety lies not in 'sacrificing hecatombs of oxen or burning countless talents of cassia', but in uncovering the Creator's wondrous works through anatomical research. Galen's anatomical descriptions thus constitute the most proper hymn to God.[16] Given this natural theology, it is obvious who Galen's opponents are: the Atomists, including the Epicureans and medical scientists such as Asclepiades of Bithynia (first century BCE). All of them subscribe to the—in Galen's eyes blasphemous—view that all bodies are conglomerates that have been formed from invisible particles in a purely mechanistic fashion.

Galen's approach to the question of the creation of the cosmos is informed by some of his most deeply held convictions about the nature and method of science. Thus it is typical of Galen to insist on the need for empirical verification. In this light, the anatomical and other evidence warrants only two assumptions: (i) that God exists; (ii) that he is good and wise. By contrast we cannot know his nature (i.e. whether or not He is corporeal or transcendent) nor the exact way in which he created the cosmos and all it contains. Indeed, it is blasphemous to lay

[15] The conjunction of Aristotelian and Platonic elements in Galen's work can be paralleled from contemporary Platonism, see Whitaker referred to above in note 10. No discrepancy was felt. Aristotle too speaks of Nature as craftsman-like, see e.g. the study by F. Solmsen, 'Nature as Craftsman in Greek Thought', *Journal of the History of Ideas*, 24 (1963), 485ff. (= *Kleine Schriften*, Hildesheim 1968, vol. 1, 344ff.), who brings out the continuity between Plato's and Aristotle's teleology. Galen also speaks of 'Nature' when referring to the creative agency. The conception of a divine, demiurgic Nature resembles the Stoic concept of God, or (universal) Nature (for which cf. note 55 below with text thereto) as an immanent creative force. But then the Stoics too were deeply influenced by the *Timaeus* and speak of their God in Platonic terms as 'demiurgic', or artistic, see e.g. Diogenes Laertius 7.134 and 137 (H. von Arnim, *Stoicorum Veterum Fragmenta*, Leipzig 1903–1924, 1.85 and 2.526).

[16] Galen, *On the Usefulness of Parts* III.10, vol. I, pp. 174.4ff., ed. Helmreich.

claim to knowledge of these subjects.[17] They belong to those issues that are insoluble because empirical testing is impossible.

It is precisely such speculative theorems over which philosophers and physicians are divided and which have given rise to competing schools or sects. The members of a school typically accept dogmas on the authority of its founder or head. And they cling to their dogmas all the more stubbornly for it. Galen deplores these scholastic divisions and the sectarian behaviour that goes with it. A true scientist does not appeal to any authority, not even great cultural heroes such as Hippocrates and Plato. Many of their views can be accepted only because they have been proved (basically) correct, not because it is they who have put them forward. So Galen may be an admirer of Plato, but he is not a member of the Platonist (or any other) school. His stance is, at least officially, individualistic. Accordingly, he likes to present his relation to the great thinkers of the classical past as direct, that is to say, unmediated by their self-styled followers in the schools.[18] In practice, to be sure, Galen was not conversing with a few other great minds in an intellectual vacuum; his readings were conditioned by his extensive philosophical education and contemporary themes and debates. In studying his work we should therefore take full account of the scholastic philosophy of his time.

The Platonic *Timaeus* had become the subject of a long exegetical tradition. Pronouncing upon one of its set issues, Galen holds that its account of creation is meant literally (see also Dillon, this volume, §1), i.e. the world has really been created, an interpretation he shared with a few (near-)contemporaries such as Plutarch and Atticus. He himself tended to treat this issue as insoluble, though he argued against Aristotle that the world's everlastingness did not preclude its having a beginning.[19] Galen, as we have noticed, believes that we cannot know the details of the process of creation. But he avoids conflict

[17] See esp. Galen, *On the Doctrines of Hippocrates and Plato*, book IX, chaps. 7 and 9, with M. Frede, 'Galen's Theology', in: J. Barnes and J. Jouanna (eds), *Galien et la philosophie* (Entretiens sur l'Antiquité classique 49), Vandoeuvres-Genève 2003, 73–130, esp. 79ff.

[18] For Galen's attitude to arguments from authority see e.g. *On the Doctrines of Hippocrates and Plato* II.3.8–11; II.4.3–4, III.8.35; V.7.83–84, ed. De Lacy. For the epistemological motivation behind his independence of the existing philosophical and medical schools of his day see the pioneering study by M. Frede, 'On Galen's Epistemology', in: V. Nutton (ed.), *Galen: Problems and Prospects*, Cambridge 1981, 65–86.

[19] See M. Baltes, *Die Weltentstehung des Platonischen Timaios nach den antiken Interpreten*, Leiden 1976, vol. 1, 63–66.

with Plato by stressing that the latter offers his account with a clear proviso: Galen is one among few ancient interpreters who gave full weight to Plato's qualification of his cosmogony as a 'likely story' (εἰκότα μῦθον, 29D2; cf. 59C6, 68D2) or 'likely account' (λόγον ... εἰκότα, 30B7; cf. 48D2) as opposed to an exact and properly scientific one. Moreover, he took it to be significant that it is Timaeus who delivers it instead of Plato's usual mouthpiece, Socrates.[20] On the other hand, this is not an ordinary Greek creation myth such as offered most notably by Hesiod's *Theogony*, with its loose ('paratactic') narrative structure (see also Bremmer, this volume, §§1.2 and 3). The Demiurge consistently makes rational choices in selecting his materials and shaping them, e.g. in the long section on the formation of man and other living things (69A6–92C9). Apart from the antiquated (and often playful) details, this is an *argument* explaining why things are as they are.[21] A reason is given even for the Demiurge's decision to create the world: being good, he wishes everything to become as much as possible like himself (29E–30A). In sum, what we have here is something rather different from traditional Greek mythology, for which Plato, followed by Galen, had little time.[22]

Galen was not a systematic theologian. Even in the *On the Usefulness of Parts* he is not clear on certain issues nor, apparently, keen to find definitive answers on them—which comes as no surprise from someone who constantly asks for empirical corroboration. His outlook left room for personal religion, which in his case was related to the cults of his native Pergamon, most notably that of Asclepius. I cannot go more deeply into this aspect of his outlook, which is of little relevance to our present purposes.[23] It is far more important to keep in mind the model provided by the *Timaeus* as well as Galen's rejection of the principle of authority and his insistence on scientific demonstration when we take a closer look at his reading of the first chapters of Genesis.

---

[20] See esp. *On the Doctrines of Hippocrates and Plato* IX.9.2–6, ed. De Lacy; *On My Own Opinions* 13.7, pp. 108.11–110.3, ed. Nutton; cf. Proclus, *In Platonis Timaeum commentaria* (ed. Diehl), vol. 1, pp. 340.21–341.9 (= Albinus, frg. 14; Gaius, frg. 9, ed. Gioè); and Apuleius, *Apologia*, chap. 12.

[21] At 51B6–7 Timaeus is explicit about his intention of providing a rational argument (λόγον).

[22] See esp. *On the Doctrines of Hippocrates and Plato* III.8.27–28, ed. De Lacy; cf. *ibidem*, 33 (quoting Plato, *Phaedrus* 229D3-E4), VI.8.78 and III.8.36.

[23] See F. Kudlien, 'Galen's Religious Belief', in: Nutton, *Galen: Problems and Prospects*, 120–121; Frede, 'Galen's Theology', 106–107.

3. *Galen on Moses on creation*

The passage that concerns us most is to be found in chap. 14 of Book XI, where Galen discusses the eyelashes, which, too, reveal the hand of God:

> Has then our Demiurge commanded only these hairs to preserve always the same length and do they preserve it as they have been ordered because they fear their master's command, or reverence the God who gave this order, or themselves believe it better to do this? Is not this Moses' way of treating Nature and is it not superior to that of Epicurus? The best way, of course, is to follow neither of these but to maintain like Moses the principle of the Demiurge as the principle (or cause, ἀρχήν) of generation of all created things, while adding the material principle to it. For our Demiurge created it to preserve a constant length because this was better. When he had decided that, he should make it [scil. the hair of the eyelashes] of such kind [i.e. of constant length and number], he set under part of it a hard body as a kind of cartilage,[24] and under another part a hard skin attached to the cartilage through the eyebrows. It was, then, certainly not sufficient merely to will their becoming thus; for had he wished to make a man out of a stone in an instant, it would not have been possible for him either. It is precisely this point in which our own opinion and that of Plato and the other Greeks who follow the right method in science differs from the position taken by Moses. For the latter it seems enough to say that God simply willed the arrangement of matter and instantaneously it was arranged; for he believes everything to be possible with God, even should he wish to make a bull or a horse out of ashes. We however do not hold this; we say that certain things are impossible by nature and that God does not even attempt such things at all but that he chooses the best out of the possibilities of becoming.
>
> In sum, we say therefore that since it was better that the length and number of the hairs of the eyelids should always remain the same, it was not that He just willed it and they were made so; for even if he just wills numberless times, they would never become of such kind [scil. as we see they are] out of a soft skin; and in particular it was altogether impossible for them to stand erect unless fixed on something hard. We therefore say that God is the cause of both things, viz. of the choice of what is better in the manufactured beings themselves and of the selection of the matter. For since it was required, first that the eyelashes should stand erect and secondly remain of the same length and number, He planted them firmly in a cartilaginous body. If He had planted them in a soft and fleshy substance, he would have suffered a worse failure not only than Moses but also than a bad general who plants a wall or a camp in marshy ground (transl. Walzer & M.T. May, modified).

---

[24] I.e. the tarsus.

At the beginning of this passage Galen deals a blow at Epicurus by comparing his approach with that of Moses. Epicurus is exposed as the inferior thinker of the two insofar as he omits the Creator from his cosmogony (see also Dillon's reply to Van Woudenberg, this volume).[25] After all, he appeals to the atoms and their random movements only, i.e. his theory is purely materialistic. Using Moses as a stick with which to beat Epicurus does not imply sympathy, let alone admiration, for the former. Indeed the formulation of Moses' position in the first sentence, I take it, is intended to brand it as flawed. The picture of the hairs fearing and obeying God's commands is the kind of personification Galen rejects in traditional mythology. But what is the explanatory model proposed by Galen?

The basic schema underlying this passage is that of the Platonic *Timaeus* interpreted in terms of the four Aristotelian causes—a reading initiated by Aristotle himself and continued by his associate Theophrastus and others.[26] Not only do we have Galen's requirement of a material cause but the Demiurge is identified with the effective cause. The form or 'due order' or 'arrangement' of the organs that the Demiurge has in mind represents the formal cause. In this particular, he decides upon the length and number of the eyelashes.[27] The final cause is represented by Galen's repeated point that the Demiurge makes his arrangements with a view to what is *better* or chooses *the best* out of the possibilities of becoming. In fact, the form and material constitution of organs such as the eye is determined by this functional and teleological perspective: the eyelashes, in contrast with hair on other places, are so formed as to facilitate our eyesight.

The repeated reference to God giving orders or simply willing certain things to come about echoes the opening chapter of Genesis, on the creation of the cosmos as a whole.[28] The use of the adverb 'instan-

---

[25] Galen, *On the Differences between Pulses* III.3 (VIII, p. 657, ed. Kühn; = Reference No. 3 Walzer) uses a similar ploy, saying 'One might more easily teach novelties to the followers of Moses ands Christ than to the physicians and philosophers who cling fast to their schools', i.e. the latter are even more sectarian in their attitude than the Jews and Christians.

[26] Aristotle, *Physica* Δ 2.209b10ff.; *De generatione et corruptione* B 1.329a9ff.; and Theophrastus, *Physicorum opiniones*, frg. 9 (ed. Diels).

[27] Note that Galen typically omits mention of the eternal and transcendent Forms, upon which the Demiurge models his creation according to the Platonic text, e.g. 30C2–31A1 and 50B5–51B6.

[28] This feature of the Mosaic account has also been preserved in the following quote from Galen's *On Hippocrates' Anatomy* by Ibn al-Matran, *Life of Galen* (vol. I, p. 77,

taneously' may also preserve a feature of chapter 1 (1:3: 'And God said: "Let there be light"; and there was light'; instantaneousness is also implied by 1:7, 9, 15). The view given to Moses is that 'God simply willed *the arrangement of matter*'. This would be a very peculiar way to refer to *creatio ex nihilo* indeed. In fact, this phrasing implies the pre-existence of matter in the Mosaic account rather than its absence.

Galen's point is that Moses omitted to specify the material selected by God because the kind of material is completely indifferent. This is what Galen means when he says that we have to add the material principle. This he himself proceeds to do when he specifies the particular material out of which the eyelashes have been formed, explaining how exquisitely suited to their function it is. Galen explains Moses' omission by reference to the latter's belief that 'everything is possible with God'—an echo of scriptural passages and, as we shall see, a current argument employed by Christians (see p. 138 below).[29] That is to say, God, being all-powerful, creates anything out of anything if he so chooses. Galen's Platonic-style God, by contrast, is limited in his power by the material with which he is confronted: '... certain things are not possible by nature and [...] God does not attempt such things at all'. However, as Galen stresses, his wisdom and benevolence are revealed by God's choice of 'the best of the possibilities of becoming'[30]—witness his selection of materials to be used for the eyes and other organs. It is worth noting that Galen does not put God above Nature but virtually identifies them by stating that God does not act against Nature.[31]

That Galen is not thinking of creation without any material is further borne out by his examples. Moses is said to take the view that God through a simple act of will could make a man out of stone. A little further on we find a reference to another substance, ash, from which the God of the Bible is able to create bulls and horses. This finds no parallel in the creation story in Genesis, or at any rate no exact

---

ed. Müller; = Reference No. 1 Walzer): 'They compare those who practise medicine without scientific knowledge to Moses, who framed laws for the tribe of the Jews, since it is his method in his books to write without offering proofs, saying "God commanded, God spake"'. The complaint that no proof is provided is frequently made by Galen (though note that he ascribes the comparison between lack of method and Moses' approach to anonymous others).

[29] E.g. Job 43:2, Mark 10:27, Matthew 19:26, Luke 18:27.

[30] Platonic language: the realm of becoming is the material world, see Plato, *Timaeus* 27D–28A.

[31] See note 15 above.

parallel. At 2:7 it is said that God made man 'of dust of the ground.' This passage comes from a section of the story following that of the creation of the world as a whole. Here *creatio ex nihilo* is no longer at issue anyway. Presently I shall suggest a few other sources of inspiration for Galen's choice of examples. For the moment we may conclude that Walzer's view that according to Galen Moses 'virtually postulated the *creatio ex nihilo* to which later Christians, Muhammadans and Jews explicitly adhered' (p. 26) misses Galen's point. His addition of the adverb 'virtually' is telling since it clearly smooths over the difficulty that the text does not really bear out his conclusion.

This having been said, it is necessary to find a more plausible explanation of at first sight unrelated elements such as echoes from Genesis 1–2 and miraculous transformations of lifeless things into living creatures. This conjunction of motives suggests that the passage may not be based on a personal or at least recent inspection of the text of Genesis. Other pagan authors from roughly the same period also refer to it (for Numenius see Van den Berg, this volume), although it is not easy to decide whether or not these references prove first-hand acquaintance with (a Greek version of) the text.[32] For the information he supplies Galen may be drawing on memory—common practice for ancient authors—, or it may even have been hearsay. One can hear other biblical echoes in his remark that God may create a man out of a stone, viz. the parallel passages Luke 3:8 and Matthew 3:9 (derived from source Q; hence no parallel in Mark). Here, as elsewhere, stone clearly stands for the lifeless.[33] A few further passages refer to stones being miraculously converted into humans or something life-dispensing such as bread.[34]

---

32 Ps.-Longinus, *On the Sublime* 9.9 with Norden; cf. also D.A. Russell (ed., intro., and comm.), *Longinus: On the Sublime*, Oxford 1964, *ad loc.*; and Numenius (favourably) *apud* Porphyry, *De antro nympharum* 10 (p. 63.10, ed. Nauck) = Numenius, frg. 30 (ed. Des Places).

33 Luke 3:8 (Jesus addressing the multitudes): 'And do not begin to say to yourselves: "We have Abraham as our father"; for I tell you, God is able from these stones to raise up children to Abraham.' Matthew 3:9: '... and do not presume to say to yourselves, "We have Abraham as our father"; for I tell you, God is able from these stones to raise up children to Abraham.'

34 Luke 19:39–40 and 4:3. Of course there is some more biblical material concerning the miraculous transformation of lifeless matter into living beings: at the Lord's command Aaron cast down his rod before the Pharaoh and his servants, whereupon it became a living serpent (the magicians of Egypt perform the same trick, but their serpents are devoured by Aaron's).

The idea of the *first* humans being created out of stone can be paralleled from Greek mythology. After the flooding of the earth Deucalion and Pyrrha created mankind anew by throwing—at the bidding of Zeus—stones over their heads: those thrown by Deucalion became men, those thrown by Pyrrha women.[35] Here too stones symbolize lifeless matter, but the context is more apposite, dealing as it does with the creation of mankind. This does not preclude the possibility that scriptural echoes are involved as well. But if Galen alludes to such myths as that of Deucalion and Pyrrha, he effectively intimates to his predominantly pagan readers the mythical nature of the Mosaic account.[36] I do not believe that Galen's example of man being created out of stone is inspired by any philosophical source.[37]

To the best of my knowledge, there are no scriptural parallels for ashes as the stuff out of which bulls or horses are created.[38] But clearly we are not dealing with an arbitrary example of an unsuitable or negligible material either.[39] The ash, or ashes (τέφρα), to a Greek imme-

---

[35] See e.g. Apollodorus, *Library*, I, vii.2.

[36] Cf. also the proverbial phrase 'being οὐ απὸ δρύος οὐδ ἀπὸ πέτρης' ('not stemming from oak or rock') in the sense of 'having parents and a country' (e.g. Homer, *Odyssey* 19.163), a phrase that also refers to mythical origins. On Galen's intended audience, see also note 14 above with text thereto. Cf. also note 22 with text thereto.

[37] *Pace* Grant, *Miracle* (note 5 above), 130, who compares Aristotle's recurrent formula 'man makes man' (e.g. *Physica* B 1.193b8, 11), as expressing the regularity of natural generation. However, the Aristotelian formula precludes the idea of creation while implying the view that species are eternal—a view for which Aristotle was criticized by Galen, see p. 130 above. Grant fails to produce an Aristotelian text containing the idea of 'man out of stone', nor have I myself been able to trace one, though notoriously Aristotle accepted cases of spontaneous generation in the animal world: see *De generatione et corruptione* Γ 11.762a5–763a30. Grant, *Miracle*, p. 130, adduces two passages from Galen's contemporaries Marcus Aurelius (viii.46) and Maximus of Tyre (XVI.4 = X.4 in the modern standard edition by Koniaris) of whom Marcus is concerned not with generation but with things typically happening to men, vines, bulls or stones, while Maximus points to the regularity of natural procreation: 'man from man (Aristotle's phrase), ox from ox, olive tree from olive tree and vine from vine'. But the context is entirely different (Platonic-style intellectual procreation). But Grant, *Miracle*, 130 is right that Galen is not concerned with *creatio ex nihilo*.

[38] Grant, *Miracle*, 130: 'In the Peripatetic treatise *De spiritu* we find the horse and the bull', viz. in chap. 9. This sounds promising, but its (pseudo-Aristotelian) author is merely saying that if the flesh and the bones of the ox and the horse differ, so must their constitutive elements. So this is not a parallel.

[39] Philo of Alexandria, *De specialibus legibus* 1.265 mentions ash (τέφρα) alongside water as the substances out of which man has been formed—a recognition which should induce modesty in us. This may certainly reflect the way Gen 2:7 was more often read.

diately suggested what is left of the funeral pyre and hence death.[40] The example therefore is suggestive of revival and rebirth. It may not be too far-fetched to recall the myth of the phoenix, the bird which periodically went up in flames and was reborn from its own ashes. Its extraordinary genesis was accepted by many ancients as a biological reality.[41] Christian authors used it as a symbol, if not proof, of the possibility of resurrection.[42] Galen may therefore be alluding here to Christians who appealed the phoenix in the context of this debate. In that case the implication is that the Judaeo-Christian view of creation entails that ordinary animals such as horses and bulls can be created in a way no less miraculous than the rebirth of the phoenix.

It is apposite to compare Walzer's Reference No. 6 (p. 15). This passage—preserved in Arabic quotations—is from Galen's lost *Summary of Plato's Republic* (see Appendix). It corresponds to the closing section of Plato's work, the eschatological 'myth' of Er (*Republic*, book X, 614B–621D; see also Van Kooten, this volume, note 55).[43] According to this story, a Pamphylian warrior named Er had been killed in battle and was lying on the funeral pyre when he came to life again. Because of his exceptional experience he was able to give an eyewitness account of the rewards and punishments awaiting human souls after death. Galen argued that the function of this Platonic myth was to offer moral guidance to those unable to follow demonstrative argument. In this connection he referred to the Christians' belief in 'parables' (and, one Arabic source adds, miracles) and to the moral benefit they derived from such tales, in particular their courage in the face of death. Clearly Galen is thinking of the resurrection in particular.

It is a fair assumption then that Galen in the passage from the *On the Usefulness of Parts* too refers to the resurrection. In the next section I shall address the question what this has to do with the creation of the cosmos with which the passage opens. For the moment we may conclude that Galen brings out the fundamental difference between the

---

40 See H.G. Liddell, R. Scott, and H.S. Jones, *A Greek-English Lexicon*, Oxford 1940 (With a Supplement, 1968), s.v.

41 Even a diehard Sceptic like Aenesidemus (first century BCE) appealed to it, see Diogenes Laertius 9.79 with R. van den Broek, *The Myth of the Phoenix*, Leiden 1972, 395–396.

42 See e.g. Clement of Rome, *Epistula ad Corinthios* (= I Clement), 1:25–26; for further evidence and discussion Van den Broek, *Phoenix*, 3ff. and *passim*.

43 Likewise Walzer, *Galen on Jews and Christians*, 57 on the authority of the Christian theological writer Ibn Zurʿa (d. 1008).

Mosaic and Platonic accounts: first, the former is defective in regard to causal theory, notably its lacking specification of a rationally acceptable material cause; secondly, it is flawed from a methodological point of view: it is a mere myth we are called upon to accept on the authority of Moses where proof is needed (we have seen how Galen tried to obviate the obvious rejoinder that the Platonic account in the *Timaeus* is no less a myth, see above, p. 131).

4. *The intellectual backdrop: some Platonist and Stoic parallels*

The attitude adopted by Galen towards Judaeo-Christian beliefs was more widely shared. Already Walzer adduced parallels from more or less contemporary Platonist literature, although he preferred to explain Galen's position by reference to the allegedly deviant Stoic Posidonius.

The first passage to be considered here[44] is from Celsus' *True Account* (Ἀληθὴς λογός) preserved by Origen in his *Against Celsus* (c. 248 CE). Celsus had completed his treatise between 177 and 180, i.e. shortly after the publication of Galen's *On the Usefulness of Parts*.[45] The context of this passage is Celsus' attack on the Jews' supposition that they are God's elect people:

> It is foolish of them also to suppose that, when God applies the fire (like a cook!), all the rest of mankind will be thoroughly roasted and that they alone will survive, not merely those who are alive at the time but also those long dead who will rise up from the earth possessing the same bodies as before. This is simply the hope of worms. For what sort of human soul would have any further desire for a body that has rotted? The fact that this doctrine is not shared by some of you [Jews] and by some Christians shows its utter repulsiveness, and that it is both revolting and impossible. For what sort of body, after being entirely corrupted, could return to its original nature and that same condition which it had before it was dissolved? As they have nothing to say in reply, they escape to a most outrageous refuge by saying that 'anything is possible to God'. But indeed neither can God do what is shameful nor does He desire what is contrary to nature. If you were to desire something abominable in your wickedness, not even God would be able to do this, and you ought not to believe at all that your desire will be fulfilled. For God is not the author of sinful desire or of disorderly confusion, but of what is naturally just and right. For the soul He might be able to provide an everlasting life;

[44] With my reading cf. Walzer, *Galen on Jews and Christians*, 30ff.

[45] See H. Chadwick, *Origen: Contra Celsum*, Cambridge 1953, xxvi–xxviii.

> but as Heraclitus says, 'corpses ought to be thrown away as worse than dung'.[46] As for the flesh, which is full of things which it is not even nice to mention, God would neither desire nor be able to make it everlasting contrary to reason. For He Himself is the reason of everything that exists; therefore he is not able to do anything contrary to reason or to his own character (5.14; transl. H. Chadwick).

Celsus counts as a Platonist, or at any rate an author with Platonist leanings, which makes his philosophical background similar to Galen's.[47] Indeed, his attitude and the terms in which it is expressed closely resemble Galen's. Celsus, like Galen, militates against the assumption that God restores the dead body. For a Platonist the body belongs to the world of becoming; it remains caught in the cycle of generation and decay. Like any decent Platonist, he reserves the eternal life for the transcendent soul. Against the Christian plea that anything is possible to God Celsus argues, in terms similar to Galen's, that God cannot and will not suspend the laws of nature.[48] From other sources we know that the Christian apologists almost standardly voiced this plea in rebutting pagan criticism of miracles and in particular the resurrection of the flesh.[49]

The passage from the *On the Usefulness of Parts* combines motifs from the Jewish-Christian account of creation with examples that seem to pertain to the resurrection. But the passage is not garbled or incoherent, as that from Celsus helps us to see. The passage on bodily resurrection in Celsus clearly belong to the same debate as that in which Galen takes part. In this debate resurrection is considered under the heading of creation. Christian sources shore up this assumption. The resurrec-

[46] Cf. Heraclitus, frg. 22 B 86 (ed. Diels-Kranz).

[47] On Celsus' philosophical affiliations, see Chadwick, *Origen*, xxiv–xxvi ('an eclectic Platonist'); cf. C. Andresen, *Logos und Nomos: Die Polemik des Kelsos wider das Christentum*, Berlin 1955, 3ff., 44ff. On Galen's philosophical position, see note 15 above with text thereto.

[48] Cf. also *Contra Celsum* 2.77 and 3.70.

[49] See Clement of Rome 27:2; Justin Martyr, *Apologies* 1.19.2; Athenagoras, *De resurrectione* 9; Irenaeus, *Adversus haereres* 5.3.2–3; and Tertullian, *De carnis resurrectione* 57 with H. Chadwick, 'Origen, Celsus, and the Resurrection of the Body', *Harvard Theological Review* 41 (1948) 83–102, 84. About a century later Porphyry, like Galen and Celsus, heaped scorn on this rejoinder, see *Against the Christians*, frg. 94 (ed. A. von Harnack), with Chadwick, 'Origen, Celsus and the Resurrection', 89–90. In fact Origen substituted the dogma of the resurrection of the flesh for a spiritualizing doctrine. Accordingly he treats Celsus' criticism in the quoted passage as beside the point. In countering Celsus' criticism of the plea 'anything is possible to God' he explains what Christians mean by it in terms very similar to what Galen says about the relation between God and nature in our passage from the *On the Usefulness of Parts*, see *Contra Celsum* 3.70.

tion of the dead marks the beginning of an entire new world. Thus the New Testament speaks of the restoration of *all* things (Acts 3:19–21).[50] In sum, the resurrection is part of a new creation by God.[51]

But where does this leave us with respect to the role played by Galen? Was he a witness to a more general reaction to Christianity or more active and influential? Walzer attached historical importance to Galen's testimonies as showing us 'the first pagan author who implicitly places Greek philosophy and the Christian religion on the same footing' (p. 43). This may be so, but Walzer, as was customary in his time, did not consider Galen to have been original. Here Walzer gave a crucial role to Posidonius (c. 135–55 BCE), whom he described as 'the great Stoic philosopher, who as a true Aristotelian united universal science and a wide metaphysical outlook in his all-embracing mind' (p. 41). In Walzer's day Posidonius was still considered not only a independent-minded, and indeed dissident, Stoic (note the qualification 'true Aristotelian'), but also an ubiquitous, if often unacknowledged, source for a wide range of extant philosophical and scientific treatises from later antiquity. It is Galen himself who tells us that Posidonius loved the truth more than his (Stoic) school, i.e. did not bow to authority (Galen, *De libris propriis* [ed. von Müller, *Scripta Minora*, vol. 2, p. 77]). Galen in *On the Doctrines of Hippocrates and Plato*, books IV–V presents Posidonius as courageous enough to abandon the school doctrine on the soul and its emotions (cf. Walzer, *Galen on Jews and Christians*, 30 note 2, 68). There can be no doubt that Galen was familiar with at least part of Posidonius' work. But what does this entail?

Walzer argued that Galen took his rejection of authority and insistence on scientific demonstration from Posidonius (*Galen*, 41–42). Fur-

[50] Cf. also Origen, *In Jeremiam homiliae* 14.18; and *Commentarium in evangelium Matthaei* 13.2 and 17.19.

[51] For the debate on the resurrection see Chadwick, 'Origen, Celsus, and the Resurrection' and Walzer, *Galen on Jews and Christians*, 30ff. J. Mansfeld, 'Resurrection Added: The interpretatio christiana of a Stoic doctrine', *Vigiliae Christianae* 37 (1983) 218–233 (= J. Mansfeld, *Studies in later Greek Philosophy and Gnosticism*, London 1989, chap. 2), focuses on the Christian appeal to the Stoic doctrine of eternal recurrence. This, too, Christian apologists argued, involved the return of individual persons such as Socrates and his accusers. A further point of contact was provided by the fact that according to the Stoic doctrine the transition to the new world was marked by the conflagration of the old one. Cf. e.g. the reference to the fire as divine instrument of punishment in Celsus' report on the Judaeo-Christian view (but it also features as an instrument of purification of sinners who can be helped). That Stoic ideas on the end of times influenced Christian thought is also argued by G.H. van Kooten, *Cosmic Christology in Paul and the Pauline School*, Tübingen 2003, 103–107. See also Adams, this volume, *passim*.

ther he argued that Posidonius had attacked the book of Genesis on lines similar to Galen's criticism. He produced a passage from Strabo's *Geography* on the nature of Jewish public life:

> Life in the state is based on ordinance; ordinance is twofold, either from gods or from men ... The ancients paid greater honour to divine command and revered it more ... For whatever its truth was, it was believed by the people and law (ἐνενόμιστο). For this very reason the seers (μάντεις) were reverenced ... since they made the divine commands and regulations public
> (xvi.2.38–39 = *FGrH* 87 frg. 70; not in Edelstein-Kidd).

Pointing to Posidonius' interest in things Jewish, Walzer (*Galen*, 21–22) took him to be Strabo's source for the above report, just he also looms behind other passages in the same work (which mention his name). Further he submits that it was Posidonius who introduced the first chapter of Genesis into the philosophical tradition (p. 30) and that his refutation of it became part of 'Middle Platonist anti-Stoic teaching' (30 note 2).[52] This refutation is simply reiterated by Galen. On the linking of anti-Jewish with anti-Stoic polemic I shall say a few things below.

Clearly Walzer's conclusions about the historical relations involved here are rash and speculative. Posidonius' broad historical and geographical interest included Jewish belief and custom (though Flavius Josephus considered him a malicious gossip).[53] But Strabo does not refer to Posidonius in the passage at issue, which is why it is omitted from the standard fragment-collection of Edelstein-Kidd (1972). The role played in Judaism by the divine commandments may have been common knowledge among intellectuals such as Strabo. But even if we were prepared to accept Posidonius as its source, these general observations could hardly count as a significant parallel to Galen's argument in *On the Usefulness of Parts*. It is hard not to see Walzer, at least on this point of his argument, as one of the latest victims of the scholarly rage called Pan-Posidonianism. By now it belongs to a past phase of scholarship. But in its heyday explicit attribution was not required when it came to sighting Posidonius behind some text.

An odd twist in Walzer's argument concerns the role played by (non-Posidonian) Stoicism in the conflict between the Judaeo-Christian

[52] Walzer added—it is only fair to report—that 'the passage of the *De usu partium* is an outcome of Galen's personal contacts with eminent Roman Jews (...) cannot be ruled out completely' (p. 30 note 2).

[53] See Josephus, *Contra Apionem* 2.79–80, 89, 91–96 (*FGrH* 87 frg. 69; Edelstein-Kidd frg. 278).

monotheistic religion and Greek philosophy. Because the Stoic concept of matter is entirely passive, the divine active principle can be said to be all-powerful: it does not meet the resistance of a pre-existing factor the way Plato's Demiurge does. Moreover, it penetrates the cosmos down to the smallest detail. Indeed, Cicero provides evidence that at least some Stoics were prepared to defend miracle-stories about divine omens and signs with an appeal to God's omnipotence.[54] Philo and later Christians such as Lactantius pointed to the omnipotent Stoic God in support of the biblical conception. Walzer appears to suppose that Judaeo-Christian thought and Stoicism really concur with respect to God's power. He takes this supposition to be supported by Posidonius alleged criticism of both Judaism and the other Stoics. Posidonius, he says, distanced himself 'from exaggerated Stoic or Jewish conceptions of the omnipotence of God' (*ibidem*, pp. 30, 32). The truth is, however, that Stoic (including Posidonian) theology is far more closer to the position expounded by Galen in the *On the Usefulness of Parts* than the Judaeo-Christian conception involving occasional interruptions of natural law and, later on, *creatio ex nihilo*. The Stoics, as is well known, actually identified God with Nature.[55] Omens and related phenomena follow from God's rational, predetermined plan and should not be taken to go against nature. So the appeal by certain Stoics to divine omnipotence as reported by Cicero means something quite different from what Galen's Jews and Christians mean by it. Indeed, when Galen and Celsus argue that God never diverges from Nature, this point owes more to Stoicism than to any other philosophical school.[56]

---

[54] Cicero, *On Divination* 2.86; and *On the Nature of the Gods* 3.92, with Chadwick, 'Origen, Celsus and the Resurrection', 85–86.

[55] See e.g. H. von Arnim, *Stoicorum Veterum Fragmenta*, Leipzig 1903–1924, 1.176, 2.945 (p. 273, lines 25–26) and 2.1024; cf. note 15 above.

[56] Walzer's argument concerning Stoicism and Posidonius' *Alleingang* lurks behind muddled statements in Grant's monograph such as the following: '... Posidonius also said that the "substance" of God was the cosmos and explicitly denied the possibility of *creatio ex nihilo* [Diogenes Laertius 7.148; H. Diels, *Doxographi graeci*, Berlin 1879, p. 462.14]. His emphasis on omnipotence, therefore, is a limited emphasis. He cannot reject science in favour of theology' (Grant, *Miracle*, 128–129). All positions ascribed by Grant to Posidonius were general Stoic doctrine and the denial of *creatio ex nihilo* common to all philosophical schools. As a Stoic, moreover, Posidonius faced no dilemma between science and theology because theology was part (indeed the crowning part) of natural philosophy.

5. *Epilogue*

Since the appearance of Walzer's book Posidonius and Galen have met with rather different reappraisals by historians of ancient philosophy. Posidonius is no longer considered all that important as a source or for that matter as a philosopher in his own right. Most historians no longer believe that he was the dissident Stoic portrayed by Galen.[57] Galen himself is no longer treated as a mere source for the ideas of other thinkers standardly taken to have been more interesting and original than he was. This rehabilitation does justice to the ancient verdict on his stature and influence.[58] It also changes the way we look at his observations on Jews and Christians. As we have seen, his methodology made him sensitive to distinctive features of the biblical theology such as the nature of divine omnipotence and the role of authoritative revelation. Yet he did not refer to *creatio ex nihilo* as such a feature. Rather, as we have noticed, he associated creation with the resurrection in a way that can be paralleled from Christian sources. I do not consider it unlikely that his typical insistence on scientific demonstration and rejection of authority influenced other participants in the debate. Eusebius, who lived about a century later but drew on an earlier source, presumably Hippolytus of Rome (d. c. 235 CE), reports that Galen even persuaded some Christians and was greatly admired by them:

> [These Christians] 'do not inquire what the holy scriptures say but sedulously endeavour to discover a form of argument to support their own ungodliness. If anyone propounds a passage of holy scripture to them they try to see whether it can produce a conditional or a disjunctive form of argument; and abandoning the sacred scriptures of God, they study geometry.' [These heretics read Euclid, they admire Aristotle and Theophrastus—and] 'Galen, I suppose, by some of them is actually worshipped' (Eusebius, *Historia ecclesiastica* 28.13–14).[59]

---

[57] For discussion and further references see T. Tieleman, *Chrysippus' On Affections: Reconstruction and Interpretation*, Leiden 2003, chap. 6.

[58] Galen was treated as a philosopher of note, see Frede, 'On Galen's Epistemology', 85–86; V. Nutton, 'Galen in the Eyes of His contemporaries', *Bulletin of the History of Medicine* 58 (1984) 315–323; cf. Tieleman, 'Plotinus on the Seat of the Soul'.

[59] On this passage see also Walzer, *Galen on Jews and Christians*, 75–86, J. Barnes, 'Galen on Logic and Therapy', in: F. Kudlien and R.J. Durling (eds), *Galen's Method of Healing*, Leiden 1991, 54. In his extant works Galen often commends the three authorities of the classical past mentioned here. So he presumably brought them to the attention of these logically minded Christians.

And we need not doubt that Galen's critique is also echoed in the opening chapter of the *Evangelical Preparation*, which is also cited by Walzer (p. 54), who however designates Porphyry's *Against the Christians* as its source:[60]

> Some have supposed that Christianity has no reason (λόγος) to support itself but that those who desire the name confirm their opinion by an unreasoning faith and an assent without examination; and they assert that no one is able by clear demonstration to furnish evidence of the truth of what is promised, but that they require their converts to adhere to faith alone, and the reason why they are called the faithful lies in their uncritical and untested faith (transl. Walzer).

As we have seen, some parallels from Celsus and other Platonist authors indicate that Galen in part used Platonist arguments. But the emphasis on demonstration (involving logic) and the firm rejection of faith in authority represent recognizably Galenic touches. Galen was not merely a witness to the debate but an active and, it seems, influential participant.[61]

APPENDIX

The Text of Walzer's References (see also footnotes 2 and 3 above)

Reference 1
They compare those who practice medicine without scientific knowledge to Moses, who framed laws for the tribe of the Jews, since it is his method in his books to write without offering proofs, saying 'God commanded, God spake' (transl. Walzer).

Reference 2
See above p. 132.

Reference 3
One might more easily teach novelties to the followers of Moses and Christ than to the physicians and philosophers who cling fast to their schools (transl. Walzer).

---

[60] If correct, this does not rule out Galen as the ultimate source with Porphyry being intermediate.

[61] I should like to thank George van Kooten for drawing my attention to a few biblical passages that are directly relevant to some of the things I wish to argue. The responsibility for the outcome is mine alone.

Reference 4
… in order that one should not at the very beginning, as if one had come into the school of Moses and Christ, hear talk of undemonstrated laws, and that where it is least appropriate (transl. Walzer).

Reference 5
If I had in mind people who taught their pupils in the same way as the followers of Moses and Christ teach theirs—for they order them to accept everything on faith—I should not have given you a definition (transl. Walzer).

Reference 6
Most people are unable to follow any demonstrative argument consecutively; hence they need parables, and benefit from them—and he [Galen] understands by parables tales of rewards and punishments in a future life—just as now we see the people called Christians drawing their faith from parables [and miracles], and yet sometimes acting in the same way [as those who philosophize]. For their contempt of death [and of its sequel] is patent to us every day, and likewise their restraint in cohabitation. For they include men but also women who refrain from cohabiting all through their lives; and they number also individuals who, in self-discipline and self-control in matters of food and drink, and in their keen pursuit of justice, have attained a pitch not inferior to that of genuine philosophers (transl. Walzer).

PART III

CREATION IN THE JOHANNINE, PETRINE, AND PAULINE LITERATURES

THE 'TRUE LIGHT WHICH ENLIGHTENS EVERYONE' (*JOHN* 1:9): JOHN, *GENESIS*, THE PLATONIC NOTION OF THE 'TRUE, NOETIC LIGHT,' AND THE ALLEGORY OF THE CAVE IN PLATO'S *REPUBLIC*

George H. van Kooten

Introduction[1]

It has long been noted by scholars that the opening of the Prologue to John's Gospel runs parallel to the opening of *Genesis*. John's well-known statement that 'in the beginning was the Word and the Word was with God, and the Word was God' (1:1) resembles and summarizes the choice of words in *Genesis*: 'In the beginning God made the heaven and the earth (...) and God said ...' (1:1–3a). This speaking of God is now rendered abstract and conceptualized as the activity of God's Word, his Logos. Plenty of attention has been paid to the Graeco-Roman background of this conceptualization. Generally, this concept of divine Logos has been understood as a Stoic notion, though it is in fact attested in ancient philosophy at large, whether in Stoic, Middle Platonist or other traditions.

However, the similarities between John's Prologue and the start of *Genesis* do not end here. Less well known, perhaps, is the fact that John also draws on what *Genesis* tells about the light, and this issue will be the central focus in this paper.[2] According to *Genesis*, 'God said: Let there be

[1] I gratefully acknowledge and thank the participants of the TBN conference for their constructive criticism and suggestions, in particular Prof. J. Dillon, who suggested I should elaborate on the availability of Greek *paideia* to Jews, including those in Palestine. Furthermore, I profited much from the discussion of this paper with Prof. M. Frede (Oxford) in which he underlined the importance of the Platonic doctrine of the 'double Helios.' I am also very grateful for comments received at the 2003 British New Testament Conference at Birmingham, as well as for those made by Prof. R. Roukema (Kampen). I wish to thank Dr Maria Sherwood Smith (Leiden) for her corrections to the English of this paper.

[2] On the congruity of the light imagery of John's Prologue and the beginning of *Genesis*, see also M. Endo, *Creation and Christology: A Study on the Johannine Prologue in the Light of Early Jewish Creation Accounts* (Wissenschaftliche Untersuchungen zum Neuen

light, and there was light. And God saw the light that it was good, and God divided between the light and the darkness' (1:3b–4). John, having dwelled for a moment on the creation by the divine Logos, continues by remarking that 'in this Logos was life, and that life was the light of mankind:' τὸ φῶς τῶν ἀνθρώπων. 'This light,' John continues, 'shines in the darkness, and the darkness has not seized it' (1:4–5).

What I shall argue in this paper is that John's interpretation of the opening of *Genesis* involves a particular Greek-philosophical understanding of light, which is as important for the understanding of his Gospel as is his notion of Logos. Maybe it reveals even more of the Graeco-Roman atmosphere in which the Gospel was written. In the first part of the paper I comment on John's view on light in his Prologue. In the second part I inquire into its function in the Gospel which follows. Together, these issues will show us the scope and content of John's interpretation of the light which God had created.

1. *The true light*

According to John, the light inherent in the divine Logos was the light of mankind: τὸ φῶς τῶν ἀνθρώπων. Soon he makes clear what he has in mind. The 'light of mankind,' which shines in the darkness, is paraphrased as 'the true light (τὸ φῶς τὸ ἀληθινόν) which gives light to everyone' and which, at the Logos' incarnation, entered into the cosmos (1:9).

It is noteworthy that the light's own activities are presented in the present tense: the light *shines* in the darkness (1:5), the true light *enlightens* everyone (1:9). This is in marked contrast with other verb groups in the Gospel's Prologue, most of which are in the past tense, since the Prologue refers almost exclusively to the past time of creation, incarnation, and Jesus' earthly ministry. The verbs describing the light's activities are meaningful exceptions. Now, as before, ever since the world's creation, the light shines in the darkness. Now, as then, the true light gives light

Testament [WUNT] II.149), Tübingen 2002, chap. 7.1.4, 217–219, esp. 219: 'As for the use of light imagery in the Johannine prologue, first of all, it reminds the readers of the event of the giving of light in the Genesis creation account'; and P. Borgen, 'Logos was the True Light: Contributions to the Interpretation of the Prologue of John,' *Novum Testamentum* 14 (1972) 115–130, esp. 117: 'the Prologue's basic structure (...) is an exposition of Gen. i 1ff.' This means that the light—darkness dualism is already part and parcel of John's use of the *Genesis* account.

to everyone. It did so already before the incarnation, the only difference being that at its incarnation the Logos-Light not only illuminated the world from without, but also entered and descended into it. But even after it has ascended again to the heavens, it still remains the true light which gives light to everyone, as it did before its descent into the world.

This will prove to be an important perspective for the readers of John's Gospel. Their present time is dominated by the presence of this all-embracing light. The difference between the time before and after the temporary dwelling of the Logos-Light on earth is not that the true light *began* to illuminate mankind, but that at least some people—the Johannine Christians—have now *become aware* of its existence and operation. Previously, mankind was ignorant of it. After the light's creation the darkness has not been able to seize it. John seems to be deliberately ambiguous here: the darkness did not grasp it, did not lay hold of it, nor did it grasp it with the mind; it did not comprehend it (1:5).

Generally speaking, this ignorance did not change after the temporary dwelling of the Logos-Light on earth. 'It was in the world; but the world, though it owed its being to it, did not recognize it,' John says (1:10). Despite the descent of the Logos-Light, even then the world at large did not recognize it, as it had not grasped it before. But apparently, Johannine Christianity is the exception. Its adherents have recognized that the true light gives light to everyone—not just to those who belong to the select group of the Johannine Christians, but to every human being. What makes a human being into a Johannine Christian is his recognition of the true light's radiation. But of what nature is this radiation? This question is not particularly difficult to answer, as the concept of true light is clearly defined in Graeco-Roman thought.

1.1. *The Greek conception of true light*

The concept of true light in John's Prologue can be traced back to Plato's *Phaedo*. In this dialogue, Socrates says:

> If someone could reach to the summit, or put on wings and fly aloft, when he put up his head he would see the world above, just as fishes see our world when they put up their heads out of the sea. And if his nature were able to bear the sight, he would recognize that that is the true heaven and the true light (τὸ ἀληθινὸν φῶς) and the true earth (*Phaedo* 109E).[3]

[3] In this paper, passages from classical literature are quoted after the standard

This distinction between the earth which lies beneath the heavens, on the one hand, and the true earth and heaven and the true light, on the other, is the same distinction as Plato makes in the *Timaeus*. In the prelude to his account, Timaeus differentiates between the visible cosmos and the invisible paradigms after which God, its architect, constructed it. The cosmos has been constructed after the pattern of that which is apprehensible by reason and thought. This visible cosmos is in fact a copy (εἰκῶν) of an invisible paradigm (παραδείγμα) which underlies it (*Timaeus* 28C–29D).

This distinction between the paradigmatic reality and its visible copy is similar to that between the true heaven and the visible heaven, the true earth and the visible earth, and between the true light and the visible light. In his *Timaeus*, Plato does not comment on this true light; he speaks only of the fire that God lighted, the sun, to give light to the whole of the visible heaven (39B), in whose light all created animals are brought out (91D), and which interacts with the light which is inherent in the eye (45B–C; 46B).[4]

As we shall see in due course, it is in his *Republic* that Plato elaborates on the qualities of that light which, in his *Phaedo*, he calls 'the true light.' In the entire ensuing Platonic tradition, this true light, the ἀληθινὸν φῶς, is also known as the intellectual light, the νοερὸν φῶς, or—alternatively—as the mental light, the νοητὸν φῶς, the light which falls in the province of νοῦς, as opposed to the visible, aesthetic light.[5] This Platonic tradition will now be examined in more detail, as against this background John's assertion that Christ is the true light which gives light to everyone gains much relief.[6]

English translations, most notably those available in the Loeb Classical Library series (Cambridge, Massachusetts: Harvard University Press) and, as far as Plato is concerned, E. Hamilton and H. Cairns (eds), *The Collected Dialogues of Plato*, Princeton, New Jersey 1999[17], with small modifications when necessary. Early patristic literature is quoted after the *Ante-Nicene Fathers* series (Edinburgh: T. & T. Clark). The New Testament is normally quoted in the Revised English Bible translation.

[4] Cf. W. Beierwaltes, *Lux Intelligibilis: Untersuchung zur Lichtmetaphysik der Griechen* (Inaugural-Dissertation München), München 1957, 38–43.

[5] Although later, in Neoplatonic thought, e.g. in Hermias, *In Platonis Phaedrum scholia* 152, there is a differentiation between hypercosmic light, noeric light and noetic light; cf. Damascius, *De principiis* 1.81.

[6] The shared occurrence of ἀληθινὸν φῶς ('the true light') in *John* 1:9 and Plato's *Phaedo* 109E seems to have gone unnoticed. Endo, for instance, in his recent study on the Johannine Prologue, is silent on the Platonic terminology of John's light imagery and refers instead to Isaiah's messianic light imagery. See Endo, *Creation and Christology*, chap. 7.2.1, 219–220; cf. chap. 8.2.3, 244–245. Peder Borgen equally neglects the

1.2. *The Platonic interpretation of the true light*

The first two authors which are particularly useful are Philo in the first half of the first century AD, and Clement of Alexandria at the end of the second, between whom—chronologically speaking—John is nicely positioned. As Philo and Clement are Jewish and Christian, respectively, we already get an impression how Jews and Christians could interpret the *Genesis* story in a Greek context.

a. *Philo of Alexandria*

In his interpretation of *Genesis*, Philo combined thoughts from both Plato's *Phaedo* and *Timaeus* (see also Dillon, this volume, §2). According

Platonic background and accounts for the term ἀληθινὸν φῶς by establishing a sharp contrast in *John* 1:8–9 between Jesus as the true, genuine, actual light and John the Baptist as the supposed, preparatory light; see P. Borgen, 'The Gospel of John and Hellenism: Some Observations,' in: R.A. Culpepper and C.C. Black (eds), *Exploring the Gospel of John*, Louisville, Kentucky 1996, 115–130, esp. 122. The occurrence of ἀληθινὸν φῶς in Plato's *Phaedo* 109E is briefly mentioned in O. Schwankl, *Licht und Finsternis: Ein metaphorisches Paradigma in den johanneischen Schriften* (Herders Biblische Studien 5), Freiburg 1995, 67, but is not integrated into Schwankl's treatment of the 'true light' in *John* 1:9 on pages 131–133. The only exception to this scholarly neglect of the Platonic-philosophical background of the 'true light' seems to be C.H. Dodd, who draws parallels between this concept in John and similar concepts in Philo and the *Corpus Hermeticum*. See C.H. Dodd, *The Interpretation of the Fourth Gospel*, Cambridge 1953, 34–35 and 50–51 (John and the *Corpus Hermeticum*, although still without references to the νοητὸν φῶς in *Corpus Hermeticum* 13.18 and the φῶς νοερόν in *Fragmenta varia* 23 [ed. A.D. Nock and A.J. Festugière 1954]), 55–56 and 203 (John and Philo), esp. 203 on John: 'His equivalent for Philo's φωτὸς ἀρχέτυπον is φῶς ἀληθινόν (i.9); both are speaking of the eternal "idea" of light, of which all empirical lights are transient copies.' Although Dodd does not refer to Plato's *Phaedo* 109E, he is well aware of the Platonic background to ἀληθινὸν φῶς in *John* 1:9, which he approaches from the angle of the epithet ἀληθινός: ''Αληθινός properly means "real" (...). Similarly, he (John) uses the term φῶς ἀληθινόν. We may then recall that Plato, in a passage which had immense influence on religious thought, offered the sun as a symbol or image of the ultimate reality, the Idea of Good, and in his allegory of the Cave suggested that as artificial light is to the light of the sun (which relative to it is αὐτὸ τὸ φῶς), so is the sun itself to the ultimate reality (*Rep.* 506D–517A). It was probably largely through the influence of Plato (...) that the conception of God Himself as the archetypal Light won currency in the religious world of Hellenism. (...) I do not suggest that the evangelist had direct acquaintance with the Platonic doctrine of Ideas; but there is ample evidence that in thoughtful religious circles at the time, and circles with which Johannine thought has demonstrable affinities, that doctrine had entered into the texture of thought. In any religious philosophy the conception of a κόσμος νοητός in some form or other was assumed—the conception of a world of invisible realities of which the visible world is a copy. It seems clear that the evangelist assumes a similar philosophy. His φῶς ἀληθινόν is the archetypal light, αὐτὸ τὸ φῶς, of which every visible light in this world is a μίμημα or symbol' (pp. 139–140).

to Philo in his writing *On the Creation* (29–36; 53), first the Maker made an incorporeal heaven (οὐρανὸς ἀσώματος), an invisible earth (γῆ ἀόρατος) and the incorporeal substance of light (ἀσώματος οὐσία φωτός). This light was an incorporeal and mental paradigm of the sun and of other heavenly luminaries: an ἀσώματον καὶ νοητὸν παράδειγμα. God says this light is beautiful for the very reason that, as a mental, intelligible light which is discernible by the mind, it surpasses the visible in the brilliancy of its radiance.

It is noteworthy that Philo already links up this concept of the intelligible light to the other important concept, that of Logos, as John does. According to John, in the Logos was life, and that life was the true light which gives light to everyone (1:4, 9). In Philo's view, too, the intelligible light is closely related to the Logos. The invisible, intelligible light came into being as an image (εἰκῶν) of the divine Logos. Together with the entire invisible cosmos, the invisible light can be said to have been firmly settled in the divine Logos (*On the Creation* 36). So the visible, aesthetic cosmos became ripe for birth after the paradigm of the incorporeal. Elsewhere, Philo stresses the fact that, whereas God is light and the archetype of every light, or rather, prior to and high above every archetype, holding the position of the paradigm of the paradigm, the Logos is indeed the paradigm which contained all God's fullness—light, in fact (*On Dreams* 1.75). Philo is apparently of the opinion that the 'Logos *is* light, for if God *said* "let there be light," this was a λόγος in the sense of a saying.'[7] Given this interpretation of *Genesis*, Philo can say that the Logos, spoken as it was when God ordered the creation of light, is itself light.

The same implication seems to be drawn in John. The Logos contains the light of mankind (1:4); it is the true light which gives light to everyone (1:9). Logos and light are closely connected. Elsewhere Philo draws the conclusion that if people are unable to see the intelligible light, they have to wander for ever as they will never be able to reach the divine λογισμός, the divine reasoning power (*On Providence* 2.19).

We now return to Philo's writing *On the Creation*, and note that he says that after the kindling of the mental light, which preceded the creation of the visible sun, darkness withdrew. Darkness withdrew as an immediate result of the creation of the intellectual light. This inference is also drawn by John. The true light contained in the Logos shines in

[7] F.H. Colson and G.H. Whitaker, *Philo in Ten Volumes*, vol. 5 (Loeb Classical Library), Cambridge, Massachusetts/London 1934, 337 note *a*.

the darkness, and the darkness has not seized it (1:5). As in Philo, this true light is not the visible light, but the intellectual light created as part of the incorporeal world before the birth of the visible world, which was about to occur after the paradigm of the incorporeal. The notion of the intellectual, true light which Philo and John use is firmly rooted in the Platonic differentiation between the intellectual and visible realms.

It is noteworthy, however, that Philo and John could hardly have experienced this notion as an unfamiliar, strange idea, as already the Septuagint offered an interpretation of *Genesis* which made it susceptible to Platonic ideas about the true, incorporeal light. It was the Septuagint which translated the very first words of *Genesis* as follows: 'In the beginning God made the heaven and the earth. But the earth was invisible and unformed:' ἡ δὲ γῆ ἦν ἀόρατος καὶ ἀκατασκεύαστος (1:1–2a). The notable difference from the Hebrew is that there the earth is not called 'invisible and unformed,' but תהו ובהו (*tohu wa-bohu*): formlessness and voidness (see also Noort, this volume, §1). The Greek phrase about the invisible earth in the beginning greatly encouraged an extensive Platonizing interpretation of the creation account in *Genesis* (see also Dillon, this volume, §2). In this way, Philo and John understood the light which was created in the beginning, when there was an invisible earth, as the true, intelligible light. Below, we will reflect on the relation between this intellectual light and the visible light of the sun, but for now we are concerned wholly with the mental type of light.[8]

b. *Clement of Alexandria*

The understanding of the first light as intellectual is also encountered in Clement's analysis of the *Genesis* story. His analysis is very interesting, as he compares Greek philosophy and the so-called 'Barbarian' philosophy of the Jewish-Christian tradition. In his *Stromateis*, he is eager to show that already the Barbarian philosophy is acquainted with Plato's differentiation between the noetic, intellectual world and the aesthetic world (5.14.93). The intellectual world is of course archetypal, whereas the visible world is the image, the material representation of the immaterial paradigm. Clement too combines Plato's *Timaeus* and *Phaedo*, when he says that this paradigmatic reality consists of the

[8] For a fuller comparison between Philo and John, see D.T. Runia, *Philo in Early Christian Literature: A Survey* (Compendia Rerum Iudaicarum ad Novum Testamentum 3.3), Assen & Minneapolis 1993, chap. 4.4, 78–83.

invisible heaven, the sacred earth, and the intellectual light.[9] Clement renders the verbal similarity with the Septuagint explicit, as he subsequently points out the correspondence between Plato and Barbarian philosophy by referring to the lines quoted above from *Genesis*: 'For "in the beginning," it is said, "God made the heaven and the earth; and the earth was invisible." And it is added, "And God said, Let there be light; and there was light."'

The creation of this reality is different from the subsequent formation of visible variety. According to Clement, 'in the aesthetic cosmogony God creates a solid heaven (and what is solid is capable of being perceived by sense), and a visible earth, and a light that is seen.' In this way, Clement demonstrates the congruity between Plato and Moses.[10] It is clear that the Septuagint text with its notion of an invisible earth at the beginning gave rise to Clement's Platonic interpretation of the first light as an intellectual light. In his Jewish-Christian tradition, he shares this understanding with Philo and John.

c. *God as the true light*

It is quite extraordinary, but we even seem to have some Graeco-Roman testimony to the Jewish-Christian speculation of God being concerned with the intellectual light. According to the Roman scholar Varro, who lived just before Philo, the Chaldeans in their mysteries call the God of the Jews Ἰάω (Varro, frg. 17; ed. Cardauns), which according to Herennius Philo of Byblos (c. AD 70–160) is Phoenician for the noetic light (*FGrH* 790 frg. 7).[11] Apparently, also among Greeks

---

[9] Cf., with explicit reference to Plato's *Phaedo*, Origen, *Against Celsus* 7.31. Having referred to Celsus' explanation of the true heaven and the true light in Plato's *Phaedo* (109E), Origen says: 'The very ancient doctrine of Moses and the prophets is aware that the true things all have the same name as the earthly things which are more generally given these names. For example, there is a "true light", and a "heaven" which is different from the firmament, and "the sun of righteousness" is different from the sun perceived by the senses' (transl. H. Chadwick, *Origen: 'Contra Celsum'*, Cambridge 1953).

[10] Cf. Clement of Alexandria, *Stromateis* 5.5.29, 'Pythagoras and his followers, with Plato also, and most of the other philosophers, were best acquainted with the Lawgiver, as may be concluded from their doctrine. (...) Whence the Hellenic philosophy is like the torch of wick which men kindle, artificially stealing the light from the sun. But on the proclamation of the Word all that holy light shone forth. Then in houses by night the stolen light is useful; but by day the fire blazes, and all the night is illuminated by such a sun of intellectual light.'

[11] Lydus, *De mensibus* 4.53: ὁ δὲ Ῥωμαῖος Βάρρων περὶ αὐτοῦ διαλαβών φησι παρὰ Χαλδαίοις ἐν τοῖς μυστικοῖς αὐτὸν λέγεσθαι Ἰάω ἀντὶ τοῦ φῶς νοητὸν τῇ Φοινίκων γλώσσῃ, ὥς φησιν Ἑρέννιος; see B. Cardauns, *M. Terentius Varro Antiquitates Rerum Divinarum*,

there was speculation about the Jewish God's identity with the noetic, intellectual light. It seems reasonable to surmise that this has to do with the Septuagint's rendering of the creation story.

Particular Greeks gods were also associated with the true, intellectual light. There is a hint in Plutarch that Osiris was understood as a conceptual light (*Isis and Osiris* 382C). According to Plutarch, the robe of Osiris has only one single colour like the light, because that which is primary and conceptual (τὸ πρῶτον καὶ νοητόν) is without admixture (*Isis and Osiris* 382C). In this, Osiris differs from Isis whose robes are variegated in their colours, since her power is not concerned with the conceptual, but with matter. Furthermore, Aelius Aristides, commenting on the Temple of Asclepius in Pergamum, regards Asclepius as the true light, saying: 'here in Asia was founded the hearth of Asclepius, and here friendly beacons are raised for all mankind by the god who calls men to him and holds aloft an ἀληθινὸν φῶς, a true light indeed' (*Orations* 23.15). That is no inordinate appraisal, as he portrays Asclepius as he who guides and directs the universe, saviour of the whole and guardian of what is immortal (*Orations* 42.4). Finally, Helios is characterized by Vettius Valens not only as a fiery commander, as one would expect, but also as an intellectual light: φῶς νοερὸν (*Anthologiarum* 1.4). These examples show that specific gods were identified with the true, intellectual light.[12]

1.3. *Enlightening every man*

Now the Platonic background of John's true light has been established, it is time to have a closer look at its description in John as the true light which enlightens every man. This further characterization also makes much sense in a Platonic context. Although Plato's digression on the intellectual light in his *Republic* will be discussed in detail below, let me already draw attention in passing to Plato's explicit statement that the

vol. 1: *Die Fragmente* (Akademie der Wissenschaften und der Literatur; Abhandlungen der Geistes- und Sozialwissenschaftlichen Klasse), Mainz & Wiesbaden 1976, 22, frg. 17; cf. Cardauns' commentary in vol. 2: *Kommentar*, 146.

[12] Cf. also the polemic about Jupiter and Christ in Clement of Alexandria, *Protrepticus* 10.98: '... the great Artist and Father has formed us, such a living image as man is. But your Olympian Jove, the image of an image, greatly out of harmony with truth, is the senseless work of Attic hands. For the image of God is his Word, the genuine Son of Mind, the divine Word, the archetypal light of light'.

prisoners in the cave should turn upward the vision of their souls and fix their gaze on that which sheds light on *all* (*Republic* 540A).

This depiction of the universal and unlimited radius of the noetic light shedding its light on all is frequently repeated in ancient philosophy. It will suffice for now to point to Epictetus and Iamblichus. Iamblichus stresses that the one and indivisible light of the gods is present, in an indivisible way, to *all those who are able to participate in it* (*On the Egyptian Mysteries* 1.9; 31.11–14). According to Epictetus, it is a shame that man honours Triptolemus, the one who taught the arts of agriculture to the nations, but tends to be negligent in service to God who acts as the true light:

> To Triptolemus, indeed, all men have established shrines and altars, because he gave us as food the fruits of cultivation, but to him who has discovered, and brought to light, and imparted to *all men* the truth which deals, not with mere life, but with a good life, —who among you has for that set up an altar in his honour, or dedicated a temple or a statue, or bows down to God in gratitude for him? (*Discourses* 1.4.31).

Epictetus characterizes God as he who has brought to light the real truth, and imparted it to *all men*.

This passage is particularly relevant as Epictetus not only stresses the universal scope of God's activity, but also employs the same verb as John: φωτίζειν ('to bring to light'). God has brought to light the real truth. This verb is not attested for the pre-Hellenistic period and is very much in vogue during the Empire to designate spiritual enlightenment.[13] Thus, Alcinous in his *Handbook of Platonism*, in a discussion about how we can conceive God, describes God's primal intellect as that which provides intellection to the power of intellection in the soul and intelligibility to its objects, by illuminating (φωτίζων) the truth contained in them (10.5; 165.23–26). God's activity of spiritual enlightenment is spoken of in terms of φωτίζειν, just as the true light in John is said to *enlighten* every man (1:9).[14]

---

[13] See J. Whittaker (intro., text, and comm.), *Alcinoos: Enseignement des doctrines de Platon* (Collection des Universités de France; L'Association Guillaume Budé), Paris 1990, 24 and 107 note 206. English transl.: J. Dillon (transl., intro., and comm.), *Alcinous: The Handbook of Platonism*, Oxford 1993.

[14] In his commentary on this passage, Whittaker links this passage from Alcinous to the one we have just quoted from Epictetus, but he also mentions *John* 1:9, the very text under discussion here (Whittaker, *Alcinoos*, 107 note 206). I leave aside the question whether *John* 1:9 should be translated as 'The true light which enlightens everyone was coming into the world' or, alternatively, as 'He (= the Logos) was the true light which

1.4. *The true light and the soul: how does it work?*

The Platonic concept of the true light should now be sufficiently clear. But how was it supposed to work? How was the true light thought to relate to mankind? It enlightens every man, but how did it actually fulfil its role of light of mankind? We have already looked briefly at the spiritual meaning of the verb φωτίζειν, but now a closer look will be taken at its proper function. Let us take another brief preview at Plato's parable of the cave in book VII of his *Republic*.

According to Plato, it is a matter of true philosophy when the prisoners are released from their subterranean cave with its shadows cast from the light of a fire, and ascend to the true light outside the cave. Although there would be some need for habituation, finally these prisoners would be able to look at the sun, i.e. they attain to the vision of the good; it is the good in the *intelligible* world which is the authentic source of truth and reason (*Republic* 514A–520D). As Plato had already explained earlier in his well-known Sun simile in book VI of his *Republic*, 'As the good is in the intelligible region to reason and the objects of reason, so is this (the sun) in the visible world to vision and the objects of vision' (*Republic*, book VI, 508B–C).

In Plato's *Republic* it is the task of true philosophy to release man from his bondage in the cave so that he may ascend to the true, intellectual light. This idea was widely received. Clement, for instance, warns his readers, whom he calls the 'sons of the true light' (οἱ τοῦ φωτὸς τοῦ ἀληθινοῦ υἱοὶ), not to close the door against this light, but to turn in on themselves, illuminating the eyes of the hidden man, and gazing on the truth itself (*Paedagogus* 2.9.80). This is very similar to what one reads in Plotinus' *Enneads*. In a passage on inner vision, Plotinus, in turn, encourages his readers to withdraw into themselves and look, and to

> bring light to all that is overcast, (…) until there shall shine out on you from it the godlike splendour of virtue (…). When you know that you have become this perfect work, when you are self-gathered in the purity of your being, nothing now remaining that can shatter that inner

---

enlightens everyone who comes into the world.' If the latter were correct, it would be notable that in the ancient world birth is very often explicitly described as 'the journey out of darkness into the light of the sun' (so Plutarch, frg. 157); see further Philo, *On the Special Laws* 3.119 on 'babes, who have just passed into the light and the life of human kind'; Plutarch, *Isis and Osiris* 355E on the birth of Osiris: 'The Lord of All advances to the light'; and Plato, *Protagoras* 320D and 321C; *Timaeus* 91D. My preference goes out to the first translation, which describes the light's descent into the world. This descent is already supposed to have been accomplished in *John* 1:10: It was *in* the cosmos.

> unity, nothing from without clinging to the authentic man, when you find yourself wholly true to your essential nature, wholly that only φῶς ἀληθινὸν, that only true light (...)—when you perceive that you have grown to this, you are now become very vision (*Enneads* 1.6.9).[15]

The same concern for man's unification with the true, intellectual light is exhibited in another treatise on the Good, or the One. In it Plotinus says that not here, but there, in the heavens, the soul may unite with its veritable love, God,

> not holding it in some fleshly embrace (...). (...) the soul takes another life as it draws nearer and nearer to God and gains participation in Him; thus restored it feels that the dispenser of true life is there to see (...). Thus we have all the vision that may be of Him and of ourselves; but it is of a self wrought to splendour, brimmed with the intellectual light (φωτὸς πλήρη νοητοῦ), become that very light, pure (...), raised to Godhood or, better, knowing its Godhood (*Enneads* 6.9.9; transl. S. MacKenna).

Such passages from Clement and Plotinus show that the Platonic concept of the true, intellectual light had clear educational connotations which aim at the edification of the soul. In this visible world, man is to find his way back to the true light. The rays of this light, Philo says, are visible to the mind only, pure from all defiling mixture and piercing to the furthest distance, flashing upon the eyes of the soul (*On Drunkenness* 44). This notion of eyes of the soul is also Platonic and widespread in ancient philosophy.[16] Two further examples may suffice at present.

(a) Already during their lives, according to Alcinous, the philosophical souls

> had longed for knowledge and had preferred the pursuit of it to any other thing, as being something by virtue of which, when they had purified and rekindled, as it were, 'the eye of the soul' (Plato, *Republic*, book VII, 533D), after it had been destroyed and blinded (...), they would become capable of grasping the nature of all that is rational (*Handbook of Platonism* 27.3; 180.22–28 [transl. J. Dillon]).

(b) According to Philo, the divine light opens wide the soul's eye (*On the Migration of Abraham* 39). Along these lines, Philo can say that there is abiding in the soul that most God-like and incorporeal light (*On Dreams*

[15] Translation, with small modifications, taken from S. MacKenna, *Plotinus: The Enneads*, London 1962³ (Third edition revised by B.S. Page).

[16] Cf. Beierwaltes, *Lux Intelligibilis*, 66–68; and, on the development towards Plato, J. Bremmer, *The Early Greek Concept of the Soul*, Princeton, New Jersey 1983, 40–41.

1.113). God shines around the soul, and the light of the intellectual light fills it through and through, so that indeed the shadows are driven from it by the rays which pour all around it (*On Abraham* 119). Consequently, when God, the spiritual sun, rises and shines upon the soul, the gloomy night of passions and vices is scattered (*On the Virtues* 164). For that reason, Philo reports, the Jewish Therapeutae pray at sunrise for a fine bright day, fine and bright in the true sense of the heavenly daylight which they pray may fill their minds (*On the Contemplative Life* 27).

The alternative for this philosophical life-style, as Philo makes quite clear, is darkness. In a passage which resembles John's Gospel very closely, Philo says that those who betray the honour due to the One 'have chosen darkness in preference to the brightest light and blindfolded the mind which had the power of keen vision' (*On the Special Laws* 1.54). This is very similar to Jesus' statement in his dialogue with Nicodemus to the effect that the light has come into the world, but that people preferred darkness to light (*John* 3:19–21). The wording also occurs in Plutarch's curious remarks on the Egyptians, who are said to have deified the field-mouse because of its blindness, since they regarded darkness as superior to light (*Table-talk* 670B). These passages clearly suggest that the road to spiritual enlightenment is not chosen automatically. Elsewhere Philo writes that some people continue to wander for ever and are never able to reach the divine reasoning power, because they are unable to see the νοητὸν φῶς, the intellectual light: the bad have lost the use of their mind, over which folly has shed profound darkness (*On Providence* 2.19). This is in marked contrast with what Philo says about others, in whose soul there is abiding that most God-like and incorporeal light (*On Dreams* 1.113).

A similar contrast between light and darkness can be found in Plutarch's polemics against the Epicureans who prefer to 'live unknown.' In Plutarch's view this lifestyle runs contrary to man's real nature. Plutarch demonstrates this by explaining the etymology of the word 'man' (φώς) from the word 'light' (φῶς). According to him, 'some philosophers believe that the soul itself is in its substance light'. For that reason the Epicurean predilection for 'living unknown' amounts to a life turned away from the light, the life of those who cast themselves into the unknown state and wrap themselves in darkness and bury their life in an empty tomb. This life very much resembles the life of those who have lived a life of impiety and crime and whose souls are eventually thrust into a pit of darkness (*Is 'Live Unknown' a Wise Precept?* 1130A–D).

This lifestyle conflicts with man's true destiny because, as Plutarch says elsewhere, 'the soul within the body is a light and the part of it that comprehends and thinks should be ever open and clear-sighted, and should never be closed nor remain unseen' (*The Roman Questions* 281). These words constitute what one might call a Platonic educational programme: the soul should be ever open to the true light that enlightens everyone. As we shall see, this is exactly the programme of John, too.

2. *The concept applied in the body of John's Gospel after the Prologue*

The conception of the true light in John's Prologue has been set against its background in Greek-philosophical thought. Now its function in the rest of the Gospel will be traced by focussing on those passages in which it reoccurs. We shall see that in two important, extensive passages John demonstrates his understanding of Christ as the true light. These passages are located in the centre of the Gospel and constitute the climax of John's reflection on this matter. But even before that there are two passage which call for attention.

2.1. *Nathanael and Jesus' power of television*

Right at the beginning of the Gospel, after the Prologue, there is a peculiar story about Jesus making the acquaintance of his prospective disciples Andrew, Peter, and Philip. The latter then goes to Nathanael and exhorts him to join Jesus, too. As soon as Nathanael comes to Jesus, Jesus hails him as an Israelite worthy of the name, in whom there is nothing false. When Nathanael is slightly embarrassed and asks Jesus how he can know this, Jesus replies: 'I saw you under the fig tree before Philip spoke to you.' At this demonstration of Jesus' apparent power of television, Nathanael converts to Jesus (1:43–49). Curious as this story may be, within the context of John's conception of the true light is becomes less cryptic. Jesus, as the divine and true, intellectual light is in no need of visible light to see clearly. That God does not demand normal daylight for his vision because he is the true light is repeatedly stressed in Philo's writings.

According to Philo, it is mistaken to assume that God 'sees nothing but the outer world through the co-operation of the sun.' As a matter of fact, God

> surveys the unseen even before the seen, for he himself is his own light. For the eye of the Absolutely Existent needs no other light to effect perception, but he himself is the archetypal essence of which myriads of rays are the effluence, none aesthetic, but all intellectual
> (*On the Cherubim* 96–97).

For that reason, to God all things are known; he sees all things distinctly, by clearest light, even by himself (*On Flight and Finding* 136).[17] If this is taken into account, it become clear that in John's Gospel Jesus' power of television arises from his role of true, archetypal, intellectual light.

2.2. *Nicodemus*

Slightly later in the Gospel, the true light is spoken of explicitly for the first time since its mention in the Prologue. In his discourse with Nicodemus, Jesus talks about the light's descent into the world, and remarks that most people prefer darkness to light, but those who live by the truth come to the light. As we have already noted, this dichotomy between those who take heed of the true light and those who do not is an integral part of Greek philosophical theory about the true light and people's attitudes to it (see section I.4).

The right attitude of mind towards the true light is subsequently demonstrated at the centre of the Gospel, in two extensive healing stories which constitute the climax of John's reflection on the true light. One is concerned with the healing of a blind man, the other with the raising of Lazarus, and neither is paralleled in the Synoptic gospels. They demonstrate the *modus operandi* of the true light.

2.3. *The healing of the blind man*

The overall theme of the two healing stories under consideration is introduced immediately previously by Jesus' statement during his public teaching in Jerusalem that he is 'the light of the world. No follower of mine shall walk in darkness; he shall have the light of life' (8:12). The meaning of this programme is immediately demonstrated, as—after his speech—Jesus sees a man who has been blind from birth. Because this blind man will be shown to be the prototype of everyone who comes

---

[17] Cf. further Philo, *On the Unchangeableness of God* 58–59; and *On the Special Laws* 1.278–279.

to see the true light, it is no coincidence that he is called 'blind from birth:' τυφλὸς ἐκ γενετῆς (9:1).

This characterization seems to be a reminder of the distinction drawn in the Gospel's Prologue (1:1–18) between being born of God (ἐκ θεοῦ γεννηθῆναι) and being born of human stock, by the physical desire of a human father (1:12–13)—a distinction which, in the dialogue with Nicodemus, is also cast as that between being born from above (γεννηθῆναι ἄνωθεν) and being born from flesh (ἐκ τῆς σαρκὸς γεννηθῆναι; 3:3–8). Those who become children of God are born from God (1:12–13) and are no longer born from flesh, or in terms of the healing of the blind man: they are no longer born blind (9:2, 19, 20, 32). It is very probable then, that the blind man is in fact the prototype of those who become children of God.[18]

When Jesus sees the blind man after his speech in which he has declared himself the light of the world (8:12), Jesus repeats this self-designation. According to John, Jesus says:

> While I am in the world I am the light of the world. With these words he spat on the ground and made a paste with the spittle; he spread it on the man's eyes, and said to him, 'Go and wash in the pool of Siloam.' (...) The man went off and washed, and came back able to see (9:5–7).

What has been said previously at the beginning of the Gospel, in a private dialogue with Nicodemus at night, is now publicly proclaimed by Jesus in Jerusalem straight after the great autumnal festival of Tabernacles (cf. 7:2).

There are two things particularly noteworthy about this healing. First of all, although Jesus is the true, intellectual light and has just spoken of himself as the light of the world, this story clearly states that the normal vision of the blind man was restored so that he could see the physical light; he came back able to see. Only on closer scrutiny is this story revealed to be about the restoration of spiritual vision.[19] It is not just about inserting vision into blind eyes. At first hand, however,

[18] This link between 'blind from birth' and 'having never beheld the true light' is also made explicitly in the exposition of the system of the Naassenes in Hippolytus, *The Refutation of All Heresies* 5.9.19: 'But if any one, he (the Naassene) says, is blind from birth, and has never beheld the true light, "which lighteneth every man that cometh into the world", by us let him recover his sight'.

[19] Otherwise, this story would have been identical with Dio Cassius' story about the healing of a blind man by Vespasian in Alexandria in AD 70: 'Vespasian himself healed two persons, one having a withered hand, the other being blind, who had come to him because of a vision seen in dreams; he cured the one by stepping on his hand and the

Jesus, the world's *true* light, imparts *physical* light to the eyes of the blind man. This presupposes some continuity between true, intellectual light and normal physical light. That seems indeed to be the case and becomes understandable if Greek philosophical thought on this matter is taken into account. According to ancient philosophers, the continuity between true, intellectual light and physical light is not just a metaphor.

(a) According to Philo, the incorporeal and intellectual light is in fact the paradigm of the sun and of all luminaries. The invisible, intellectual light is a supercelestial constellation and at the same time the source of the constellations obvious to the senses (*On the Creation* 29–31). As a matter of fact, God, as the archetype on which laws are modelled, is the sun of the sun; he is 'the noetic of the aesthetic:' he is in the intellectual realm that which the sun is in the perceptible realm, and from invisible fountains he supplies the visible beams to the sun which our eyes behold (*On the Special Laws* 1.279).

(b) In a similar way, Plutarch is of the opinion that one must not believe that the sun is merely an image (εἰκῶν) of Jupiter, but that the sun is really Jupiter himself ἐν ὕλῃ, in his material form (*The Roman Questions* 282C).

(c) The continuity between intellectual and physical light is also stressed by Vettius Valens who calls Helios a fiery commander as well as an intellectual light: φῶς νοερὸν (*Anthologiarum* 1.4).

(d) Likewise, throughout his *Hymn to King Helios* Julian makes clear that Helios, the sun, enlightens both the intellectual and physical reality: 'For just as through his light he gives sight to our eyes, so also among the intelligible gods through his intellectual counterpart (...) he bestows on all the intellectual gods the faculty of thought and of being comprehended by thought' (145B). At the same time Helios possesses intellectual functions and a visible creative function (145D).[20]

---

other by spitting upon his eyes' (Dio Cassius 65.8.1–2; transl. E. Cary); these miracles were interpreted as a sign that 'Heaven was thus magnifying him.'

[20] On the dual function of Helios, see also W. Fauth, *Helios Megistos: Zur synkretistischen Theologie der Spätantike* (Religions in the Graeco-Roman World 125), Leiden 1995, xxxi, with reference to *Corpus Hermeticum* 16.17–18: '(Helios) bildet (...) das demiurgische Bindeglied zwischen intelligibler Welt (κόσμος νοητός) und sinnlich wahrnehmbarer Welt (κόσμος αἰσθητός), transportiert das Gute (τὸ ἀγαθόν) von oben nach unten, wobei er selbst gemäß dieser Kommunikation Mittelpunkt der kosmischen Sphären, der Kosmos hingegen das Werkzeug seiner demiurgischen Aktivität ist'; and 135–137, with reference to Proclus. For a detailed commentary of Proclus' *Hymn to Helios*, in which Helios is addressed as 'king of νοερὸν φῶς (*Hymn* 1.1),' 'king of noeric fire,' see R.M.

(e) All four examples seem to be a reflection of Plato's statement, in book VII of his *Republic*, that the idea of good 'is indeed the cause for all things of all that is right and beautiful, giving birth in the *visible* world to light and the sun ("and its lord"), and its own power in the *intelligible* world producing truth and reason' (517B–C).[21] Against this background,[22] one can more easily discern why in John's Gospel Christ, the true, intellectual light, can at the same time impart physical light to the eyes of the blind man; the true light is simultaneously the physical light of this world.

---

van den Berg, *Proclus' Hymns: Essays, Translations, Commentary* (Philosophia Antiqua 90), Leiden 2001, 153: 'the sun is characterized by a double procession from the Demiurgic Nous. In its humbler manifestation it is just one of the heavenly bodies. According to (Proclus' interpretation of) *Ti*[*maeus*] 39B4, however, the Demiurge himself gave the sun its light "not from a material substrate, but from himself." Hence it is also called "noeric light" (νοερὸν φῶς [...]). This light does two things: on the one hand it creates order and harmony in the universe (...); on the other hand it elevates all things to the Demiurgic Nous.'

[21] See also Plato's allegory of the sun in his *Republic* 507B–509C. Cf. Beierwaltes, *Lux Intelligibilis*, 37–57, esp. 51–52 on the similarity between the idea of good and the sun: 'Sonne und ἀγαθόν sind zwar voneinander verschieden, trotzdem besteht eine Übereinstimmung. Beide stimmen nicht nur darin überein, daß sie Ordnungs- und Lebensprinzip sind, sondern daß sie auch von Wesen Licht sind. Die Sonne spendet das Licht, damit der Gesichtssinn die Gegenstände des Sehens wahrnehmen kann. Das ἀγαθόν gibt den Dingen die Wahrheit (508E), daß sie erkannt werden können. Im ersten und im zweiten Bereich macht das Licht die Dinge sichtbar und einsehbar. Im ersten ist es das sinnliche Licht, im zweiten das intelligible. Das intelligible Licht ist dem sinnlichen logisch und ontologisch vorgeordnet.'

[22] See further Philo, *On the Creation* 55: 'It was with a view to that original intellectual light, which I have mentioned as belonging to the order of the incorporeal world, that He created the heavenly bodies of which our senses are aware'; *On the Migration of Abraham* 40: 'Wisdom is God's archetypal luminary and the sun is a copy and image of it.' Note also Hermias of Alexandria, *In Platonis Phaedrum scholia* 177 on the 'double Helios': 'Überall sagt Platon, daß der Herrscher Helios in Analogie zum ersten Prinzip steht. Wie nämlich hier die Sonne Herrscher über den ganzen wahrnehmbaren Kosmos ist, so ist es über den noetischen jener. Und wie von dem Herrscher Sonne Licht hinabgebracht wird, welches das Sehfähige mit dem Sichtbaren verbindet, verknüpft und eint, auf dieselbe Weise verknüpft auch das Licht, das aus dem ersten Gott hervorgeht—er nennt es "Wahrheit"—, den Nous mit dem Noetischen. Man kann also sehen, daß die Schönheit dies nachahmt. Denn sie ist gleichsam ein Licht, das ausgesendet wird von der Quelle des Noetischen hin zum irdischen Kosmos' (H. Bernard [transl. and intro.], *Hermeias von Alexandrien: Kommentar zu Platons 'Phaidros'* [Philosophische Untersuchungen 1], Tübingen 1997; cf. *In Platonis Phaedrum scholia* 179). On the issue of the dual function of the sun in relation to Christ as the 'true sun,' see M. Wallraff, *Christus Verus Sol: Sonnenverehrung und Christentum in der Spätantike* (Jahrbuch für Antike und Christentum; Ergänzungsband 32 [2001]), Münster 2001, Index, s.v. 'Doppelte Sonne' and 'Geistige Sonne.'

Secondly, it is indeed noteworthy that this healing story is not just about physical light and physical vision. As we already surmised, the blind man functions as the prototype of those who come to be born from God, born from on high, and who thus receive spiritual enlightenment. This is not only implicit in Jesus' dual identity as the light of the world, but is also rendered explicit in Jesus' remark that he has come into this world, to give sight to the sightless, but to make blind those who claim to see (9:39–40). This confirms our impression that the healing of the blind man is in fact a prototypical example of spiritual enlightenment. Soon this illustration of the true light's activity is followed by another healing story which features another prototype, who is not merely healed from blindness but is even raised from his grave in a cave.

2.4. *Lazarus*

The prototype who figures in the other healing story is Lazarus. In many respects, Lazarus is an even more powerful exemplar of life turned towards the true light than is the blind man, as he is first raised from the dead and then regains his power of sight when a cloth, wrapped around his face, is finally removed. According to John, Jesus was informed early on of the serious illness of his friend Lazarus, yet deliberately delayed his visit to him, so that he would indeed die. Jesus explains his delay by stating: 'Anyone can walk in the daytime without stumbling, because he has this world's light to see by. But if he walks after nightfall he stumbles, because the light fails him' (11:9–10).

The point Jesus apparently wants to demonstrate is that because he—the light of the world—is away from Lazarus, Lazarus is short of this light and stumbles to his death. This is what Jesus wants to make evident to the people, and for that reason, for their sake, he is even glad that he was not there (11:15). Only after Lazarus' death and funeral does Jesus arrive. The correct understanding of the whole situation, however, is about to dawn for those among the crowd who had already experienced Jesus' healing of the blind man. They ask themselves: 'Could not this man, who opened the blind man's eyes, have done something to keep Lazarus from dying?' (11:37)

The answer to this question is given by Jesus, who goes to the tomb, which is in a cave—as John explicitly says—, and orders Lazarus to come out. In response, 'the dead man came out, his hand and feet bound with linen bandages, his face (ὄψις) wrapped in a cloth. And

Jesus said, "Release him; let him go:"' λύσατε αὐτὸν καὶ ἄφετε αὐτὸν ὑπάγειν (11:38–44). Impressed by this event, many come to believe in Jesus, though the authorities now reach their definitive decision to kill Jesus and 'to do away with Lazarus as well, since on his account many Jews were going over to Jesus' (11:45–53; 12:9–11).

The prototypical value of this story of the raising of Lazarus springs to mind very easily. Again John applies the concept of true light, and this time there appear to be notable parallels with Plato's parable of the cave. This seems no coincidence, since after all John's Prologue had already explicitly introduced Jesus as the true light. This concept is derived from Plato's *Phaedo*, but is worked out in full in book VII of his *Republic*, in the well-known parable of the prisoners in the cave, who are gradually introduced to the real light of the sun outside the cave.

2.5. *Plato, Greek education, and the Jews*

Before I come to making a case for the correspondences between Plato's allegory of the cave and John's story of how Lazarus was raised from a cave by Jesus, the true light, it seems imperative above all to outline how John could have known Plato. The degree to which John, in his portrayal of Jesus, seems to be familiar with Plato's thought cannot be explained satisfactorily by a vague reference to a *Zeitgeist* in which such notions were general currency. Rather, such knowledge hints at familiarity with particular Platonic notions through some form of education (*paideia*).

John's acquaintance with Plato could be the result of formal, institutionalized education, but that is not necessary, as a whole range of formal and informal training and teaching in Greek language, culture, and philosophy was available throughout the Mediterranean world.[23] Jews had access to it, too. That they even had knowledge of Plato is clear from explicit references to him by Jews such as Aristobulus, Philo of Alexandria, Josephus, and Justus of Tiberias.

According to Aristobulus, who probably lived in Alexandria in the second century BC, Plato imitated Jewish law, which was available to him in a partial Greek translation predating the Septuagint; the philosopher had worked through each of the details contained in it,

[23] On the ubiquity of *paideia* in the Hellenistic and Roman worlds, see T. Morgan, *Literate Education in the Hellenistic and Roman Worlds*, Cambridge 1998, 3, 21–25.

and had taken many things from it (Aristobulus, frg. 4). Aristobulus is convinced that

> since Pythagoras, Socrates, and Plato investigated everything thoroughly, they seem (...) to have followed him (Moses) in saying that they hear God's voice by reflecting on the cosmic order as something carefully created by God and permanently held together by him (frg. 4).[24]

This conviction about Plato's dependence on the Jewish Scriptures, which—as we shall see—was shared by Josephus, could certainly enhance a favourable attitude towards Plato among Jews. Such congeniality is found in Philo's writings in the first half of the first century AD. In his work *On the Creation*, Philo refers to Plato with approval: '..., as Plato says, ...' (119: ὡς ἔφη Πλάτων; cf. 133). Furthermore, he refers explicitly to Plato's *Timaeus* (*On the Eternity of the World* 13; 25; 141) and seems to side with Plato in his view of the indestructibility of the cosmos (13–17; 27). He pays Plato a compliment when introducing a quotation from him: 'And so Plato says well ...' (38), and calls him ὁ μέγας Πλάτων, the great Plato (52). Although at times in his writings Philo explicitly criticizes Plato (*On the Contemplative Life* 57–59), nevertheless he does not refrain from calling him also ὁ ἱερώτατος Πλάτων, the most sacred Plato (*Every Good Man is Free* 13).

In line with this Jewish affinity with Plato is Josephus' appreciation of this philosopher. In his writing *Against Apion*, written around the turn of the first century AD, Josephus ventures historiographical views similar to those of Aristobulus, to the effect that the wisest of the Greeks learned to adopt fitting conceptions of God from principles with which Moses supplied them. Among these Greeks, Josephus also mentions Plato by name, adding that such philosophers appear to have held views concerning the nature of God which were similar to those of Moses (2.168).

Later on, Josephus even defends Plato's attempt to draft a constitution (πολιτεία) and code (νομοί) against current criticism: Plato is continually being scoffed at and held up to ridicule by those who claim to be expert statesmen (2.222–225). Interestingly, Josephus defends Plato against unjustified criticism of his *Republic*, showing Jewish acquaintance with this specific dialogue in the first century AD. It is no surprise then that Josephus further demonstrates his full sympathy with Plato

[24] Translation taken from C.R. Holladay, *Fragments from Hellenistic Jewish Authors*, vol. 3: *Aristobulus* (Texts and Translations 39; Pseudepigrapha Series 13), Atlanta, Georgia 1995.

by pointing out analogies between Plato's laws and those of the Jews, and highlights points in which Plato followed the example of Moses, the Jewish law-giver (2.256–257). In so doing, Josephus refers implicitly to Plato's *Republic*.

It seems highly relevant to our present enquiry that there is so much explicit and positive reference among Jews to Plato in the periods both immediately preceding and contemporaneous with John. Supposing that John indeed had some knowledge of Plato, the examples from Aristobulus, Philo, and Josephus illustrate that this would not have been altogether impossible or even exceptional for a Jew. A possible objection might be that Aristobulus and Philo represent the highly Hellenized Judaism of Alexandria, and that Josephus wrote his *Against Apion* in Rome, whereas the origins of John's Gospel lie in first-century Palestine.

However, modern research has argued that an imagined contrast between a non-Hellenized Palestine and a Hellenized Jewish Diaspora is unwarranted.[25] This can also be clearly shown with regard to the issue at hand, since explicit Jewish acquaintance with Plato is not restricted to the Diaspora. Diogenes Laertius, the early third-century AD author of a compendium on the lives and doctrines of ancient philosophers, mentions Justus of Tiberias as the source of an apocryphal story about Plato's intercession at Socrates' trial (*Lives of Eminent Philosophers* 2.41). Justus is known from Josephus' writings as the son of a Jewish faction leader in Tiberias (*The Life* 31–42). Tiberias was one of the chief cities of Galilee besides Sepphoris and Gabara (123), founded by Herod the Great's son Herod Antipas after the accession in 14 AD of Emperor Tiberius and named after this dignitary (*The Jewish War* 2.167–168; *Jewish Antiquities* 18.36). Tiberias not only had a Galilean-Jewish population, but also Greek residents (*Jew. Ant.* 18.37; *The Life* 67). In this Galilean city then, the Jew Justus was able to cultivate an interest in Plato.[26]

---

[25] See, e.g., M. Hengel, *Jews, Greeks and Barbarians: Aspects of the Hellenization of Judaism in the pre-Christian Period*, Philadelphia 1980, esp. chap. 12: 'The Influence of Hellenistic Civilization in Jewish Palestine down to the Maccabean Period'; J.J. Collins and G.E. Sterling (eds), *Hellenism in the Land of Israel* (Christianity and Judaism in Antiquity Series 13), Notre Dame, Indiana 2001; M. Hengel, 'Judaism and Hellenism Revisited,' in: Collins & Sterling, *Hellenism in the Land of Israel*, Notre Dame 2001, 6–37, esp. 7: 'it is misleading to distinguish fundamentally between a "Palestinian Judaism" in the motherland and "Hellenistic Judaism" in the Diaspora as is still usual.'

[26] On Justus of Tiberias, see E. Schürer, G. Vermes, F. Millar, M. Black, *The History of the Jewish People in the Age of Jesus Christ (175 BC—AD 135)*, vol. 1, Edinburgh 1973, 34–37.

As Josephus acknowledges, Justus was not unversed in Greek *paideia*: οὐδ' ἄπειρος ἦν παιδείας τῆς παρ' Ἕλλησιν (*The Life* 40). This shows that Justus had had access, in some way, to Greek learning. It implies knowledge of Greek, although not necessarily of the standards achieved by the Herodian rulers who—according to Josephus—had reached the highest degree of Greek *paideia* (*The Life* 359). Levels of proficiency in Greek will have varied. Josephus himself says that he has 'laboured strenuously to partake of the realm of Greek prose and poetry, after having gained a knowledge of Greek grammar' (*Jew. Ant.* 20.263).

But apart from knowledge of Greek, Justus must also have become familiar with philosophy, as is apparent from his interest in Plato and Socrates. This need not suggest that Justus was formally trained. Although philosophy seems to have constituted the climax of Greek *paideia* after preliminary studies (Philo, *On the Preliminary Studies* 74–76)[27] and to have been an element of formal, institutionalized education (*On the Special Laws* 2.229–230), it was also accessible through less formal channels. As Philo shows, men can also be involved in the study of philosophy from the very cradle and in a less systematic way (*On Drunkenness* 51).[28] In the Hellenistic and Roman period, Greek culture was spread by the sum total of institutions like gymnasia, palaestrae, libraries, theatres, thermae, temples, stadiums, forums, and agoras.[29] Palestine could not and did not avoid this 'global' process of Hellenization.

In Palestine, Greek culture had been a presence since Alexander the Great, and even the allegedly anti-Hellenistic revolt of the Jewish Hasmoneans (the 'Maccabees') in 168/167 BC seems to have been directed only against the excessive policy of one particular Greek-Seleucid ruler, Antiochus IV Epiphanes. The Hasmoneans themselves,

---

[27] Cf. A. Mendelson, *Secular Education in Philo of Alexandria* (Monographs of the Hebrew Union College 7), Cincinnati 1982, chap. 2.

[28] Cf. Mendelson, *Secular Education*, 44. Cf. also T. Dorandi, 'Organization and Structure of the Philosophical Schools,' in: K. Algra, J. Barnes, J. Mansfeld and M. Schofield (eds), *The Cambridge History of Hellenistic Philosophy*, chap. 3, esp. 61: 'Beside this kind of organized and institutionalized school (*scholai*, *diatribai*), there were also groups of people who got together to practise philosophy in an apparently less rigidly structured form, which could be defined as a "pseudo-school" or, better, "philosophical tendency" (*agōgai* or *haereseis*).'

[29] See the following articles in *Der neue Pauly*: I. Hadot, 'Gymnasion, II. Das Hellenistische Gymnasion,' vol. 5 (1998), 23–27; K. Vössing, 'Bibliothek, II.B Bibliothekswesen,' vol. 2 (1997), 640–647; J. Gerber and V. Binder, 'Hellenisierung,' vol. 5 (1998), 301–312; and J. Christes, 'Bildung,' vol. 2 (1997), 663–673; 'Erziehung,' vol. 4 (1998), 110–120; and 'Paideia,' vol. 9 (2000), 150–152.

as a matter of fact, took the initiative of sending diplomatic letters to Sparta (*1 Maccabees* 12:1–23; Josephus, *Jew. Ant.* 13.163–170). In this correspondence the Hasmoneans stated that they wanted to renew their ties of brotherhood (ἀδελφότης) in reply to a previous Spartan letter in which it was stressed that Spartans and Jews were brothers and that they both descended from Abraham: εἰσὶν ἀδελφοὶ καὶ (...) εἰσὶν ἐκ γένους Αβρααμ (*1 Macc* 12:21; cf. Josephus, *Jew. Ant.* 12.225–226: ἐξ ἑνὸς [...] γένους). This means that they were regarded as sharing the same συγγένεια, the same kinship (Josephus, *Jew. Ant.* 13.164; 13.170).[30] The construction of Jewish kinship with the Spartans shows that even the Hasmoneans wanted to be part of the Hellenistic world. It is just one example of the general tendency in the world of Hellenism to discover one's Greek origins and to express this in terms of kinship.[31]

A pivotal role in this ongoing process of Hellenization was played by Herod the Great and his successors, to whom the Romans granted the Hasmoneans' political power from 37 BC onwards. Herod's philhellenism led to an increase in institutions such as cities, gymnasia and theatres by which Greek culture was spread both within and without his Jewish kingdom. To the North of his territories, Herod provided gymnasia,[32] theatres, halls, porticoes, temples and agoras for cities such as

---

[30] For a positive Jewish attitude towards the Spartans, see also *1 Macc* 14:16–23; Josephus, *Jew. War* 1.513–515; *Against Apion* 2.225–227. Spartans and Cyrenians were also thought to be genetically related, according to Josephus, *Jew. War* 2.381.

[31] Cf. T. Rajak, 'Hasmoneans,' in: S. Hornblower and A. Spawforth, *The Oxford Classical Dictionary*, Oxford & New York 1996[3] (= *OCD* [3]), 668–669; S. Hornblower, 'Hellenism, Hellenization,' in: *OCD* [3], 677–679; and Hornblower, 'Kinship,' 807–808. For the importance of establishing συγγένεια as a 'passport to Greek culture,' see I. Hadot, 'Gymnasion, II. Das Hellenistische Gymnasion,' in: *Der neue Pauly*, vol. 5 (1998), 23–27, esp. 26. On this Greek practice, in which Jews participated, see O. Curty, *Les parentés légendaires entre cités grecques: catalogue raisonné des inscriptions contenant le terme suggeneia et analyse critique* (Centre de recherches d'histoire et de philologie de la IVe section de l'École pratique des hautes études III; Hautes études du monde gréco-romain 20), Genève 1995; C.P. Jones, *Kinship Diplomacy in the Ancient World* (Revealing Antiquity 12), Cambridge, Massachusetts 1999 (these monographs were kindly brought to my attention by J.N. Bremmer); and S. Lücke, *Syngeneia: epigraphisch-historische Studien zu einem Phänomen der antiken griechischen Diplomatie* (Frankfurter althistorische Beiträge 5), Frankfurt am Main 2000.

[32] Even though the installation of a Greek gymnasium at Jerusalem during the excessively anti-Jewish policy of Antiochus IV Epiphanes sparked of the Hasmonean revolt in 168/167 BC (*1 Macc* 1:14–15; *2 Macc* 4:7–12; and Josephus, *Jew. Ant.* 12.240–241), Jews as such were not against participation in gymnasia. Josephus refers to Greek-Seleucid privileges that Jews who went to the gymnasium but were unwilling to use foreign oil, out of religious scruples about purity, should receive recompensation from the gymnasiarchus (the general supervisor of the civic gymnasia) to pay for their own

Tripolis, Damascus, Ptolemais, Berytus (Beirut), Tyre and Sidon (Josephus, *Jew. War* 1.422). But within his territories too he built theatres, both in Jerusalem (Josephus, *Jew. Ant.* 15.268–280)[33] and in Caesarea, where he also built an amphitheatre and agoras (*Jew. War* 1.415; *Jew. Ant.* 15.341). According to Josephus, the theatre of Jerusalem was acceptable to most Jews, as soon as they were reassured that it contained no images which would desecrate the Holy City (*Jew. Ant.* 15.272–280). As was acknowledged in another case, even a visit to the theatre in Caesarea would not render one impure (*Jew. Ant.* 19.330–334).

This enumeration of Herod's building activities shows the vast range of his programme, which also included the foundation of a new town in Samaria, only one day's journey from Jerusalem. The new town was called Sebaste after Augustus and contained a massive temple devoted to the emperor (Josephus, *Jew. War* 1.403; *Jew. Ant.* 15.292–298).[34] This shows that Herod's philhellenism manifested itself both within and without his Jewish kingdom.

To turn back to Justus of Tiberias, the entire digression on the Hellenization of Palestine from Alexander the Great, through the Hasmonean period, right up to the Herodian-Roman age sharpens our awareness of how Justus could have become acquainted with Plato even in Galilee. As a citizen of Tiberias, a city founded by Herod Antipas and inhabited by a mixed Jewish-Galilean and Greek population, he could have learned Greek either informally or through some form of education. The remains of a large early Roman building in Tiberias have been tentatively interpreted as a palaestra or a gymnasium.[35] Justus might have encountered Platonic philosophy through a (visiting) teacher who taught in such palaestrae and gymnasia.[36] But it

kind of oil (*Jew. Ant.* 12.119–120). For further evidence of Jews participating in gymnasia, see M.H. Williams, *The Jews Among the Greeks and Romans: A Diaspora Sourcebook*, London 1998, 107: V.1–2; 112–114: V.20–24. Philo, too, seems to speak from personal experience. See Philo, *On the Special Laws* 2.229–230; cf. Mendelson, *Secular Education*, 28–33.

33 On the importance of Herod for the Hellenization of Jerusalem, see K.M. Kenyon, 'Aelia Capitolina (Jerusalem)', in: R. Stillwell (ed.), *The Princeton Encylopedia of Classical Sites*, Princeton, New Jersey, 12–13: 'Jerusalem cannot really be said to have entered the Classical world until the time of Herod the Great in the last third of the first century BC' (p. 12).

34 On Herod's philhellenic building programme, see D.W. Roller, *The Building Program of Herod the Great*, Berkeley 1998.

35 A. Negev, 'Tiberias,' in: Stillwell (ed.), *The Princeton Encyclopedia of Classical Sites*, 920–921.

36 Cf. Morgan, *Literate Education*, 29: 'Some teachers may have taught in gymnasia or palaestrae, but we are not in a position to say that they were regular places for schools.'

is the entire interface between institutions such as cities, agoras, gymnasia and theatres in the region which accounts for the transmission of Greek culture. In the theatres of both Jerusalem and Caesarea, for instance, Greek plays will have been staged.[37]

In light of this culture, John's Gospel with its Platonic concept of the 'true light' could have been written anywhere in the Palestinian area. Whether John's Gospel had its roots in the Galilean city of Tiberias (*John* 6:1, 23; 21:1), had a Samaritan connection, as the opponents' characterization of Jesus a 'a Samaritan' might suggest (8:48; cf. 4:39–40),[38] or originated in Jerusalem, Greek culture was sufficiently present in Galilee, Samaria, and Judea to account for John's Greek conceptualizations.[39] All that is needed for John's Gospel to be written is for an author like Justus of Tiberias to become a follower of Jesus. That something like this is not unthinkable, may be gleaned from Josephus, who, after expert training in the 'philosophical schools' of the Pharisees, Sadducees, and Essenes, deemed this education insufficient and became a devotee of a certain Bannus in the desert for a period of three years (*The Life* 10–12).[40]

What I suggest is that the author of John's Gospel might well have become acquainted with Plato within the context of Greek *paideia* somewhere in Palestine, just as happened in the case of Justus of Tiberias. That the author of John's Gospel became familiar with Plato's allegory

---

Although there is, remarkably enough, an 'almost complete absence, for any period of antiquity, of evidence for any kind of formal central control or organization of teachers or schools or what was taught' (Morgan, p. 25), there is ample evidence of the ubiquity of education. According to Morgan, p. 3, 'at any time from the early third century BCE until the end of the Roman empire, you could be fairly sure of finding a teacher, or more than one, in most towns and many villages, in the forum, at the crossroads, in the gymnasium, or in a private house or garden.'

[37] Cf. J.N. Bremmer, 'The Atonement in the Interaction of Greeks, Jews, and Christians,' in: J.N. Bremmer and F. García Martínez (eds), *Sacred History and Sacred Texts in Early Judaism* (Contributions to Biblical Exegesis and Theology 5), Kampen 1992, 75–93, esp. 92 and 81–82. On archaeological finds of a theatre in Jerusalem, see R. Reich and Y. Billig, 'A Group of Theatre Seats Discovered near the South-Western Corner of the Temple Mount', *Israel Exploration Journal* 50 (2000) 175–184 (I owe this reference to the kind suggestion of J.N. Bremmer).

[38] For a Samaritan background to John, see R.E. Brown, *The Community of the Beloved Disciple*, London 1979, 36–40.

[39] Greek conceptualizations in John would be even less surprising if John were written in Ephesus, as Irenaeus claims (Irenaeus, *Against Heresies* 3.1.1; cf. Eusebius, *Ecclesiastical History* 5.8.4).

[40] For a commentary on this passage, see S. Mason, *Life of Josephus: Translation and Commentary* (Flavius Josephus Translation and Commentary 9), Leiden 2001, 15–20.

of the cave in particular can be explained from the fact that in antiquity this simile in book VII of the *Republic* was well-known among Plato's works. Philo, for instance, draws on it in his criticism of contemporary sophists, and says that they,

> unable to discern the intellectual light (τὸ νοητὸν φῶς) through the weakness of the soul's eye (...) as dwellers in perpetual night disbelieve those who live in the daylight, and think that all their tales of what they have seen around them (...) are wild phantom-like inventions
> (*Every Good Man is Free* 5).

The parable or traces of it are also found in, among others, Plutarch,[41] Alcinous,[42] Iamblichus,[43] Gnostic authors,[44] and Plotinus.[45]

2.6. *Lazarus and Plato's cave*

If indeed Plato was known among Jews, even among Jews in first-century AD Galilee, as the case of Justus of Tiberias demonstrates, it is no surprise that John, too, could be familiar with him. Moreover,

---

[41] See Plutarch, *How the Young Man Should Study Poetry* 36E: 'But when they (the young men) hear the precepts of the philosophers, which go counter to such opinions, at first astonishment and confusion and amazement take hold of them, since they cannot accept or tolerate any such teaching, unless, just as if they were now to look upon the sun after having been in utter darkness, they have been made accustomed, in a reflected light, as it were, in which the dazzling rays of truth are softened by combining truth with fable, to face facts of this sort without being distressed, and not to try to get away from them' (cf. *Republic* 515E).

[42] Alcinous, *Handbook of Platonism* 27.4, 180.28–39.

[43] Iamblichus, *Protrepticus* 15–16.

[44] See the Gnostic Naassenes in Hippolytus, *The Refutation of All Heresies* 5.10.2: ποτὲ ⟨μὲν⟩ βασίλ(ειον) ἔχουσα βλέπει τὸ φῶς, ποτὲ δ' εἰς ⟨σπ⟩ήλαιον ἐκρι⟨πτο⟩μένη κλάει (ed. M. Marcovich, *Hippolytus: Refutatio Omnium Haeresium* [Patristische Texte und Studien 25], Berlin & New York 1986, 171; cf. Th. Wolbergs, *Griechische religiöse Gedichte der ersten nachchristlichen Jahrhunderte*, vol. 1: *Psalmen und Hymnen der Gnosis und des frühen Christentums* [Beiträge zur klassischen Philologie 40], Meisenheim am Glan 1971, 49–50): 'Sometimes she (the soul) would live in a royal palace and look at the light; but sometimes she is being thrown in a cave, and there she weeps' (cf. M. Marcovich, 'The Naassene Psalm in Hippolytus [*Haer.* 5.10.2],' in: B. Layton [ed.], *The Rediscovery of Gnosticism*, vol. 2: *Sethian Gnosticism* [Studies in the History of Religions 41.2], Leiden 1981, 770–778).

[45] Plotinus, *Enneads* 2.9.6: 'there is nothing here but a jargon invented to make a case for their (the Gnostics') school: all this terminology is piled up only to conceal their debt to the ancient Greek philosophy which taught, clearly and without bombast, the ascent from the cave and the gradual advance of souls to a truer and truer vision' (transl. S. MacKenna). That these Gnostics were Christian can be surmised on account of Porphyry, *On the Life of Plotinus* 16. In my view, John had already appropriated Plato's allegory of the cave.

the cave parable from book VII of Plato's *Republic* was among the best-known passages of his writings.

The following direct or inverted parallels between book VII of Plato's *Republic* and John suggest themselves. This parallelism is found either in John's Lazarus story, in the story about the blind man, or at other levels of John's Gospel. It appears impracticable to treat these levels in isolation, as various threads from the contents and context of Plato's cave parable seem to be interwoven into the Johannine fabric. To use another image, the resonances of particular Platonic themes from the cave parable make themselves heard throughout John's Gospel. For this reason, I shall go backwards and forwards between the story of Lazarus, that of the blind man, and the Gospel at large.

The two most important reasons to assume that John's Gospel echoes themes from Plato's cave parable are (1) the specific combination of 'light' (φῶς) and 'cave' (σπήλαιον), and (2) the characterization of this light as the true, non-physical light which enlightens all.

1. *The pair 'light' and 'cave'*

At the beginning of book VII of his *Republic*, Plato depicts men who dwell in a cave-like dwelling (ἐν καταγείῳ οἰκήσει σπηλαιώδει) which, over the entire width of the cave (παρὰ πᾶν τὸ σπήλαιον), is open to the light (φῶς; 514A).

This specific combination of the terms 'light' and 'cave' reoccurs later, when Socrates tells Glaucon, his discussion partner, that as part of their education the best pupils, who had once been liberated from the cave, should be sent down into the cave (σπήλαιον) again. After a fifteen-year period, they should be brought out again and required to

> turn upwards the vision of their souls and fix their gaze on that which sheds light (φῶς) on all, and when they have thus beheld the good itself they shall use it as a pattern for the right ordering of the state and the citizens and themselves throughout the remainder of their lives (539E–540A).[46]

[46] As W. Jaeger has emphasized, despite this political talk about 'the ordering of the state,' the 'ultimate interest of Plato's *Republic* is the human soul. Everything else he says about the state and its structure (...) is introduced merely to give an "enlarged image" of the soul and its structure. But even in the problem of the soul, Plato's interest is not theoretical but practical. He is a *builder of souls*. He makes Socrates move the whole state with one lever, the education which forms the soul.' See W. Jaeger, *Paideia: The Ideals of Greek Culture*, vol. 2: *In Search of the Divine Centre* (translated by G. Highet), New York & Oxford 1943, 199.

This explicit contrast between cave and light also features in John's story about Lazarus. Because Jesus, the light of this cosmos (11:9: τὸ φῶς τοῦ κόσμου τούτου), is away from Lazarus, Lazarus lacks this light (11:10: φῶς), stumbles to his death, and is buried in a cave (11:38: σπήλαιον). After Jesus has awakened him, in his final public teaching in Jerusalem, Jesus exhorts his audience to be receptive towards the light (φῶς; 12:35–36, 46) and to become children of light (12:36).

The combination of 'light' and 'cave' is a clear echo of Plato's parable.[47] The change from the 'normal' prisoners' cave of Plato's parable into the burial cave in the Lazarus story can be explained as the outcome of some further associative thought. Plotinus, too, in his retelling of Plato's parable, portrays the souls as having been *buried* in a cave: τεθάφθαι τε λέγεται καὶ ἐν σπηλαίῳ εἶναι (*Enneads* 4.8.4).[48]

---

[47] Other examples of the after-effects of Plato's cave and light imagery are probably the traditions about Jesus' birth in a cave and the cave symbolism in the cult of Mithras. See Justin, *Dialogue with Trypho* 78.6 on Jesus' birth in a cave; this is understood as a fulfillment of the prophecy of *Isaiah* 33:16 LXX: 'he shall dwell in a high cave (σπήλαιον) of a strong rock' (70.1–2), a prophecy which Justin regards to have been imitated by Mithras (70.1–2; 78.6). The cave is also mentioned in the *Protoevangelium of James* 38–39, and in Origen, *Against Celsus* 1.51 (cf. H. Chadwick, *Origen: 'Contra Celsum'*, Cambridge 1953, 47 note 5). In the *Protoevangelium of James* the birth of Jesus in a cave is accompanied by a great light: 'And they (Joseph and the midwife) went to the place of the cave, and behold, a dark (bright) cloud overshadowed the cave. And the midwife said: "My soul is magnified to-day, for my eyes have seen wonderful things; for salvation is born to Israel." And immediately the cloud disappeared from the cave, and a great light (φῶς) appeared in the cave (σπήλαιον), so that our eyes could not bear it. A short time afterwards that light withdrew until the child appeared' (19:2; transl. O. Cullmann, in: W. Schneemelcher and R.McL. Wilson [eds], *New Testament Apocrypha*, vol. 1: *Gospels and Related Writings*, Cambridge & Louisville, Kentucky 1991). See M. Gervers, 'The Iconography of the Cave in Christian and Mithraic Tradition,' in: U. Bianchi (ed.), *Mysteria Mithrae* (Études Préliminaires aux Religions Orientales dans l'Empire Romain 80), Leiden 1979, 579–599; and A. Meredith, 'Plato's "Cave" (*Republic* vii 514a–517e) in Origen, Plotinus, and Gregory of Nyssa,' in: E.A. Livingstone (ed.), *Studia Patristica* 27, Louvain 1993, 49–61. The cave of Mithras receives a Platonizing interpretation in Porphyry, *The Cave of the Nymphs in the Odyssey* 6. See R. Turcan, *Mithras Platonicus: Recherches sur l'Hellénisation philosophique de Mithra* (Études Préliminaires aux Religions Orientales dans l'Empire Romain 47), Leiden 1975, esp. 23–27, 65–67, 133; cf. also D. Ulansey, 'Mithras and the Hypercosmic Sun,' in: J.R. Hinnells (ed.), *Studies in Mithraism*, Rome 1994, 257–264.

[48] Cf. E. Hoffmann, 'Der pädagogische Gedanke in Platons Höhlengleichnis,' in: H.-Th. Johann (ed.), *Erziehung und Bildung in der heidnischen und christlichen Antike* (Wege der Forschung 377), Darmstadt 1976, 118–131, esp. 130: 'Der Weg der Erziehung (...) verlangt eine Abkehr von der natürlichen Sinnlichkeit; aber er löst dem Menschen die Fesseln, führt ihn ins freie Reich der Gedanken, ermöglicht ihm, nach übersinnlichen Gesichtspunkten die Welt zu verstehen (...). (...) Wer ihn gegangen ist, weiß nun, daß die Höhle ein Grab war'.

2. *The nature of the light*

Apart from the distinctive combination of 'cave' and 'light,' it is also the characterization of this light which points in the direction of Platonic thought. In Plato's cave parable, the ascension from the cave upwards (ἡ ἄνω ἀνάβασις) signifies the soul's ascension to the intelligible region (ἡ εἰς τὸν νοητὸν τόπον τῆς ψυχῆς ἄνοδος), and the sunlight it encounters outside the cave is emitted by the idea of good. This idea, according to Plato, is indeed the cause for all things of all that is right and beautiful, giving birth in the visible world (ἔν τε ὁρατῷ) to light (φῶς) and its author (the sun), whereas in the intelligible word (ἔν τε νοητῷ) it itself is the power of truth (ἀλήθεια) and reason (517B–C). Implicitly, Plato draws a distinction here between the physical light, which is emitted by the visible sun, and the non-physical, true, intelligible light—the distinction we have come across before and which evolves from the mention of the true light in Plato's *Phaedo*.[49] Moreover, this non-physical, intelligible light comes into view again at the end of book VII of Plato's *Republic* in the passage, already quoted, in which Socrates says that the best pupils should be required 'to turn upward the vision of their souls and fix their gaze on that which sheds light (φῶς) on all:' εἰς αὐτὸ ἀποβλέψαι τὸ πᾶσι φῶς παρέχον (540A). This light is the non-physical, intelligible, true light.

It is this light which is in view in John, too. The Lazarus story is both introduced, and its meaning reinforced, by Jesus' self-proclamation as τὸ φῶς τοῦ κόσμου τούτου (11:9–10; 12:35–36, 46), the light of this cosmos. The same holds true of the introduction to the story of the blind man (9:5; cf. 8:12). Jesus' repeated self-designation as the light of this cosmos seems to suggest a link between the two stories. The link between someone who was born blind and someone dwelling in a cave seems anything but far-fetched. Sextus Empiricus, for instance, in what seems to be an allusion to Plato's cave parable, says that those who live in subterranean and unlighted caves (οἵ τε ἐν καταγείοις τισὶ καὶ ἀλαμπέσι σπηλαίοις βιοτεύοντες) and those who are blind from birth (οἱ ἐκ γενετῆς πηροὶ) do not hold a true conception of particular things

[49] On the close thematic similarity between the ascent from the cave towards the light of the sun in Plato's *Republic* 517B and the true light in his *Phaedo* 109E, cf. Beierwaltes, *Lux Intelligibilis*, 63: 'Dieser Aufstieg aus der Höhle ist im Phaidon mythisch vorgebildet: aus den Höhlen (κοῖλα 109B5, 109C2) gelangen nur ganz reine Naturen zur Betrachtung des wahren Lichtes und der wahren Erde (τὸ ἀληθινὸν φῶς καὶ ἡ ὡς ἀληθῶς γῆ 109E). Auch hier gibt es Erkenntnisstufen, die vom dunklen Unten zum hellen Oben reichen.'

(*Against the Physicists* 2.175 [10.174]). In John, the blind-born (9:1: τυφλὸς ἐκ γενετῆς) and Lazarus seem to be connected in a similar way. Both encounter the light of the cosmos, which has been introduced in John as the true light which enlightens everyone: τὸ φῶς τὸ ἀληθινόν, ὃ φωτίζει πάντα ἄνθρωπον (1:9).

One can scarcely fail to notice the close parallel between this light that enlightens all and 'that which sheds light on all' in Plato's *Republic* (540A). Both the distinctive contrast between cave and light, and this light's identity as the true, non-physical light seem to point to John's familiarity with the simile of the cave in Plato's *Republic*.

How much else from book VII of the *Republic* resonates in John must probably remain a matter for debate. I shall discuss some other less direct, sometimes even inverted but nevertheless highly remarkable parallels. If one assumes that the direct parallels mentioned above must be the result of John's *paideia* in Greek culture, these other similarities can probably also best be explained as due to John's acquaintance with Plato's *Republic*. For the sake of clarity, I shall continue enumerating the possible points of contact between John and book VII of Plato's *Republic*. These points consist of: (3) an implicit comparison between Socrates and Jesus in John's Gospel, (4) the release from bondage in the cave, (5) the issue of 'inserting vision into blind eyes,' (6) the contents of Plato's *paideia*, and, finally, (7) the accessibility of his *paideia*.

3. *Socrates and Jesus*

To start with an 'inverted' parallel, I draw attention to the beginning of Plato's allegory of the cave. After Plato has told how one prisoner is freed from his bonds, dragged up the ascent, comes out into the light of the sun and, after a period of habituation, is able to see the things higher up (τὰ ἄνω; 516A), Plato subsequently describes what would happen to this man εἰ πάλιν ὁ τοιοῦτος καταβὰς (516E), if he were to go down again. According to Plato, he would provoke laughter among his former fellow prisoners who would be ignorant of his need to adjust again to the darkness of the cave, and would argue instead that his eyes had apparently been ruined when he had gone upwards (ὡς ἀναβὰς ἄνω), so that it would not be worthwhile even to attempt such an ascent (517A). Finally, if it were possible to kill the man who now tried to release them and lead them up, they would do so (517A).

Plato is clearly alluding here to the death of Socrates, and implies that Socrates' contemporaries did indeed kill him when he came down again: πάλιν ὁ τοιοῦτος καταβὰς. In John this action of coming down is ascribed to Jesus, as he is ὁ ἐκ τοῦ οὐρανοῦ καταβάς, the one who came down from heaven (3:13). At this point, John seems to invert the parallel between Socrates and Jesus. Whereas Socrates came down (καταβάς) into the cave after his upward ascension (ἀναβὰς ἄνω), Jesus did not ascend prior to his descent. In fact, John emphasizes, nobody ascended into heaven except the one who came down from heaven: οὐδεὶς ἀναβέβηκεν εἰς τὸν οὐρανὸν εἰ μὴ ὁ ἐκ τοῦ οὐρανοῦ καταβάς (3:13).

In this way, John inverts the parallel between Socrates and Jesus: Jesus descended without prior ascension, and Socrates did not ascend to heaven at all. Later on in book VII of the *Republic* not just Socrates, but other gifted prisoners, too, are said to be led upwards to the light (φῶς), 'even as some are said to have gone up from Hades to the gods:' ὥσπερ ἐξ Ἅιδου λέγονται δή τινες εἰς θεοὺς ἀνελθεῖν (521C). Against this background, John's polemic is easier to understand.[50] It can hardly be a coincidence that in both Plato's *Republic* and in John the language of καταβαίνειν (to descend) and ἀναβαίνειν (to ascend) is highly dominant.[51] That is not to say that John's use of it has been occasioned by Plato, but at least its application will have been further shaped by Plato's *Republic*.[52]

---

[50] Cf. Borgen, 'The Gospel of John and Hellenism,' 102–104 and 116, esp. 103: 'In different forms the idea of ascent to heaven was widespread in the wider Hellenistic world. When John reacted against persons' claims of ascent within a Jewish context, he reacted against a Jewish (and Christian) phenomenon that at the same time took place within a Hellenistic context.'

[51] Καταβαίνειν in Plato, *Republic*, book VII, in 516E, 519D, 520C; καταβαίνειν in *John* in 1:51, 3:13, 6:33, 38, 41–42, 50–51, 58. Ἀναβαίνειν in Plato, *Republic*, book VII, in 517A, 519D and ἀνάβασις in 515E, 517B, 519D; ἀναβαίνειν in *John* in 1:51, 3:13, 6:62, 20:17. In addition to this, also κάτω (*Republic*, book VII, in 519B, 527B, 529A–C; *John* 8:23), ἄνω (*Republic*, book VII, in 516A, 517A–C, 525D, 527B, 529A–C, 533D; *John* 8:23, 11:41), and ἄνωθεν (*Republic*, book VII, in 514B, 518B; *John* 3:3, 7, 31 and 19:11) occur in both writings. On the importance of ἀνάβασις in Plato's *Republic*, see K. Albert, *Griechische Religion und Platonische Philosophie*, Hamburg 1980, 50–60: 'Anabasis,' esp. 50: 'In mehreren Dialogen geht Platon auf das Thema des Aufstiegs (…) ein. (…) Der Begriff der Anabasis wird im Zusammenhang mit dem "Höhlengleichnis" der *Politeia* mehrfach von Platon verwendet (515E, 517B, 519D)' and 54: 'Die (…) zuvor angeführten Texte sind die wichtigsten, in denen Platon die philosophische Erkenntnis als ἀνάβασις, als Aufstieg versteht, auch wenn das Wort selbst nur in der *Politeia* vorkommt.'

[52] The customary references to ascents into heaven in Jewish texts (see, e.g., A.F. Segal, 'Heavenly Ascent in Hellenistic Judaism, Early Christianity and their Environment', in: W. Haase [ed.], *Aufstieg und Niedergang der römischen Welt*, vol. II.23.2, Berlin 1980, 1333–1394, esp. 1352–1368) do not seem to be sufficient, however. Ascents to heaven in Jewish literature are attributed to figures like Enoch, Abraham, Moses,

This conscious comparison between Socrates and Jesus seems to extend further. According to Plato, again alluding to Socrates, a man returning from divine contemplations to the miseries of men appears most ridiculous if, not yet accustomed to the darkness, he is compelled in courtrooms to contend about the shadows of justice (517D). This man's soul, Plato says, has come ἄνωθεν ἐκ φωτὸς, from the light above (518A–B). This description of Socrates' provenance again corresponds with John's portrayal of Jesus, who is ὁ ἄνωθεν ἐρχόμενος, he who comes from above (3:31). Both Socrates and Jesus are described as one who came down (ὁ καταβάς) from above (ἄνωθεν).

Moreover, it is not only Socrates who provokes a discussion about who is actually able to see, he who came down or those who had remained in darkness and question the usefulness of attempting to go upwards (517A). In John's story about the man who was blind from birth, a similar discussion develops between the blind man who has been cured from his blindness, the Pharisees who do not believe that the man had been blind and had now gained his sight, and Jesus, who causes offence by implying that those who claim to see are in fact themselves blind (9:13–41).

Finally, not just Socrates is killed after he has come down from the light above (517A). In John's story about Lazarus, immediately after Jesus' operating as the true light at Lazarus' cave, the Jewish Council plots to kill Jesus (11:53),[53] and to do away with Lazarus as well, because his awakening from the cave has caused many to put their faith into Jesus (12:9–11). In John, Socrates and Jesus seem to be put on a par,

---

Baruch, and Isaiah, yet, as scholars such as Meeks, Dunn and Segal acknowledge, the Johannine pattern of descent and ascent has no direct parallel in Jewish literature. See W.A. Meeks, *The Prophet-King: Moses Traditions and the Johannine Christology* (Supplements to Novum Testamentum 14), Leiden 1967, 297; J.D.G. Dunn, 'Let John be John a Gospel for Its Time', in: Peter Stuhlmacher (ed.), *Das Evangelium und die Evangelien* (WUNT I.28), Tübingen 1983, 309–339, esp. 328–329; and Segal, 'Heavenly Ascent', 1375 on the *katabasis-anabasis* pattern in John: 'This is not the first time that the complete *anabasis-katabasis* pattern has been evidenced. But in the past the complete pattern has been limited either to a presumed descent and ascent of an individual soul or to the announcement of a divine message by means of an angel who ascends after having delivered it. Only one half of the journey had any real significance.' In this respect, the parallels between John and Plato's *Republic* have at least complementary value; they share a complete *anabasis-katabasis* pattern. I wish to thank E.J.C. Tigchelaar for discussing this issue with me.

[53] According to John, there had been previous attempts by the Jews to seize and kill Jesus (*John* 5:16, 18; 7:1, 19, 25, 30; 8:37, 40), but somehow John regards the resurrection of Lazarus as the turning point in the Jews' plotting to kill Jesus (11:53).

albeit in a somewhat concealed form, only recognizable for those who know both stories. It is, however, the same inverted parallelism which comes to the fore in later Christian authors, such as Justin Martyr (cf. also Van den Berg, this volume, §4). In his *Apologies*, Justin draws parallels between Socrates and Christ, whereby they are subsequently presented as opposites.[54] What happens in John is essentialy the same.

4. *The release from bonds*

Considering the sceptical and hostile reception for Socrates after his descent, Plato asks himself rhetorically, 'And if it were possible to lay hands on and kill τὸν ἐπιχειροῦντα λύειν τε καὶ ἀνάγειν, the man who tried to release them and lead them up, would they not kill him?' (517A). Depending on whether the previous parallels have proven convincing, the following resemblance between Plato's simile of the cave and John's story about Lazarus could also be relevant.

After Jesus has appeared at Lazarus' cave as the world's true light and has awakened him,[55] Lazarus emerges from the cave with his hands and feet still bound (δεδεμένος) and with his ὄψις, his face or power of vision, still bound round (περιεδέδετο) with a cloth (11:44a). Following Lazarus' appearance, Jesus orders him to be released and permitted to go forth: Λύσατε αὐτὸν καὶ ἄφετε αὐτὸν ὑπάγειν (11:44b–c). This double command to release (λύειν) Lazarus and to let him go forth (ὑπ-άγειν) seems to mirror Socrates' double endeavour to release (λύειν) the prisoners in the cave and lead them up (ἀν-άγειν). In Plato, the phase of release from bondage (532B: λύσις τε ἀπὸ τῶν δεσμῶν; cf. 515C) is subsequently followed by conversion (μεταστροφή) and ascent (ἐπάνοδος) to the world above (532B).

[54] See Justin, *Apologies* 1.5.4 and 2.10. Cf. also Lucianus, *The Death of Peregrinus* 12 on the Greek philosopher Peregrinus (died AD 165), who after his conversion to Christianity and imprisonment was called by the Christians 'the new Socrates.'

[55] The implicit portrayal of Christianity as the true philosophy seems to have been visualized in the fourth century AD. See J. Kremer, *Lazarus: Die Geschichte einer Auferstehung. Text, Wirkungsgeschichte und Botschaft von Joh 11,1–46*, Stuttgart 1985, 160: 'auf jüngeren Darstellungen aus dem 4. Jahrhundert, besonders auf vielen Sarkophagen, trägt der Erwecker des Lazarus einen Philosophen-Mantel und hält eine Buchrolle in Händen (…): die Auferweckung des Lazarus ist gleichsam ein Bild für die erlösende Kraft des Christentums als wahrer Philosophie.' The general theme of resurrection from the dead was not altogether absent from Graeco-Roman literature. Cf. Kremer, p. 97: 'In den Schriften der griechisch-römischen Antike ist von einer Totenerweckung mehrfach die Rede.' Notably, in the tale of Er in Plato's *Republic*, there is talk of a dead man, Er, coming to life again (*Republic*, book X, 614B–621D; see also Tieleman, this volume, §3).

5. *The issue of 'inserting vision into blind eyes'*

At this point there seems to arise a notable difference between John and Plato. In Plato's *Republic*, the release from bondage is followed by a conversion from the shadows to the images that cast them and to the light (532B). Plato stresses that conversion is not a matter of inserting vision (ὄψις) into τυφλοῖς ὀφθαλμοῖς (blind eyes); rather, what is needed is the 'conversion (περιαγωγή) of the soul, not an art of producing vision in it, but on the assumption that it possesses vision but does not rightly direct it and does not look where it should, an art of bringing this about' (518B–D). Conversion is about redirecting one's eyes and power of vision, not about inserting vision into blind eyes, as if vision were not already existent. This seems to differ greatly from John's story about the blind man, whose eyes were blind and had to be opened (9:1–7). His restored vision is contrasted with the (mental) blindness of the Pharisees (9:39–41). In this respect, John and Plato do conflict, as Plato stresses the pre-existence of vision, even though it is in need of redirection by *paideia*.

Yet, even Plato is not entirely consistent in his application of the imagery of eye-sight. In book VII of his *Republic*, he also speaks about the fact that the soul's instrument of knowledge needs to be purified and kindled afresh by *paideia* because it has been destroyed and blinded (τυφλούμενον) by the ordinary habits of life (527D–E). Ignoring his earlier criticism of viewing *paideia* as the insertion of vision into blind eyes (518B–D), Plato himself slips into the common imagery of mental blindness.

At the same time, John's concept of the power of vision might be more subtle than it first appears. John's story of Lazarus seems to suggest that after Lazarus' awakening, his power of vision (ὄψις) was existent but needed to be uncovered (11:44). But even if one considers this interpretation too far-fetched, and accepts that John and Plato do indeed differ to some extent, it is nevertheless undeniable that there is some similarity in their figurative use of blindness, even though this mention of mental blindness is exceptional in Plato and, philosophically speaking, incorrect.

6. *The contents of Plato's paideia*

Besides the similarity between John and Plato's *Republic* with regard to the light and cave imagery, there are also striking resemblances in John with the *paideia* which, according to Plato, leads towards the light. Conversion, in Plato's view, entails the soul's turning (μεταστροφή) from the shadows to the images that cast them and to the light (532B), from the world of generation to the truth (525C: μεταστροφὴ ἀπὸ γενέσεως ἐπ' ἀλήθειάν). This terminology of 'generation' and 'truth' is pivotal in both John and Plato. The soul must be turned around from the world of becoming (518C: ἐκ τοῦ γιγνομένου) and be cast free of the leaden weight of birth and becoming (γένεσις), which attach themselves to the soul by food and similar pleasures and gluttonies and turn down the vision of the soul (519A–B).

The same stress on the deficiency of the natural world as such is characteristic of John. It is not sufficient that one is born in the natural sense of physical generation (1:12–13); it is also necessary to be born from above: Δεῖ ὑμᾶς γεννηθῆναι ἄνωθεν (3:3, 7). This need is exemplified in the story of the man τυφλὸς ἐκ γενετῆς, blind from the hour of birth (9:1). For the same reason, Jesus exhorts his audience to long for the true bread (6:32: ὁ ἄρτος ὁ ἀληθινός), the real food (6:55: ἀληθής βρῶσις) and real drink (ἀληθής πόσις), and not to strive after the food that passes away, but rather the food that lasts, the eternal food: ἐργάζεσθε μὴ τὴν βρῶσιν τὴν ἀπολλυμένην ἀλλὰ τὴν βρῶσιν τὴν μένουσαν εἰς ζωὴν αἰώνιον (6:27).[56] This concords with Plato's criticism of food that turns down the vision of the soul (519A–B) and with his recommendation of knowledge of that which always is, and not of a something which at some time comes into being and passes away: τοῦ ἀεὶ ὄντος γνῶσις, ἀλλὰ οὐ τοῦ ποτέ τι γιγνομένου καὶ ἀπολλυμένου (527B).

Conversion, in Plato's view, is not only turning away from the perishable world of generation, but also, positively, turning towards the truth: μεταστροφὴ ἀπὸ γενέσεως ἐπ' ἀλήθειάν (525C). In both Plato and John, ἀλήθεια (truth) is a key term and seems to be closely connected with φῶς (light), implying an etymological wordplay on ἀλήθεια, which is understood as ἀ-λήθεια, i.e. 'unconcealedness,' truth, and reality. According to Plato, after his ascent from the cave, the former prisoner is drawn into the light (φῶς), but at first unable to see even one of τὰ ἀληθῆ (515E–516A), the things that are real.[57] Light and truth

[56] Cf. also the dialogue with the Samaritan woman on the difference between normal water and living water in *John* 4:13. Just as the Samaritan woman asks Jesus: Κύριε, δός μοι τοῦτο τὸ ὕδωρ (4:15), thus in his dialogue about the true bread the audience asks him: Κύριε, πάντοτε δὸς ἡμῖν τὸν ἄρτον τοῦτον (6:34). Just as normal water does not stop one from becoming thirsty again (4:13), normal bread nourishes one only for a limited period (6:26).

[57] The same etymological understanding of ἀλήθεια and its link with φῶς underlies

are closely related, because what light is in the visible word, truth is in the intelligible world (517C). Those who convert towards τὰ ἀληθῆ, the things that are unconcealed, real and true (519B), experience a turning around from a nightly day to the true, veritable day (521C: περιαγωγὴ ἐκ νυκτερινῆς τινος ἡμέρας εἰς ἀληθινήν) and lead the βίος τῆς ἀληθινῆς φιλοσοφίας, the life of true philosophy (521B; 521C).

The same interest in truth and its association with light is exhibited in John. It seems no coincidence that they occur first together as a compound expression, when Christ is called τὸ φῶς τὸ ἀληθινόν (1:9), the unconcealed, true, real light. Using the wording of Plato's *Phaedo* (109E) to distinguish the true light from the physical light, John further implements the distinction between 'true' (ἀληθινός) and 'physical' by talking, for example, of the truth (ἀλήθεια) generated by Christ (1:14, 17), the readiness of those who pursue the truth (ἀλήθεια) to come to the light (φῶς; 3:21), the true worshippers (οἱ ἀληθινοὶ προσκυνηταὶ) who worship in spirit and truth (4:23), the true bread (ὁ ἄρτος ὁ ἀληθινός; 6:32), the true food (ἀληθής βρῶσις; 6:55) and the true drink (ἀληθής πόσις) as opposed to perishable food.

It is no surprise that in John's Gospel this interest in truth culminates in Pontius Pilate's question: Τί ἐστιν ἀλήθεια; (18:38), 'What is truth?' It seems probable that John conceived the answer to this question in terms of Christ's identity as τὸ φῶς τὸ ἀληθινόν, the world's unconcealed, true, real light (1:9). In comparison with the Synoptic gospels, the language of truth is frequent and intense in John and this seems to be grounded in the notion of the true light. The close association between 'light' and 'truth' in John seems to reflect a concern which is very similar to the paideutic enterprise of book VII of Plato's *Republic*.[58]

In Plato, the conversion to the light (532B) and towards truth (519B; 525C; 527B) is also expressed by means of a contrast between 'upwards' or 'on high' (ἄνω) on the one hand, and 'downwards' or 'below' (κάτω)

---

also Philo, *On Joseph* 68: φῶς γὰρ ἡ ἀλήθεια ('truth is light'); Plutarch, *The E at Delphi* 387A: 'Philosophy is concerned with truth, and the illumination of truth (ἀληθείας φῶς) is demonstration;' and Alcinous, *Handbook of Platonism* 10.3 (164.38–40): God is 'the truth (ἀλήθεια), because he is the origin of all truth, as the sun is of all light (φῶς).'

[58] If this is true, Rudolf Bultmann's well-known characterizaton of the Johannine Jesus becomes obsolete: 'Thema seiner Rede is immer nur das Eine, daß der Vater ihn gesandt hat, daß er gekommen ist (…), daß er wieder gehen wird (…). So zeigt sich schließlich, das Jesus als der Offenbarer Gottes nichts offenbart, als daß er der Offenbarer ist (…). Johannes stellt also in seinem Evangelium nur das Daß der Offenbarung dar, ohne ihr Was zu veranschaulichen.' See R. Bultmann, *Theologie des Neuen Testaments*, Tübingen 1958[3], 414, 418–419.

on the other. According to Plato, the conversion to τὰ ἀληθῆ, the things which are real, redirects the vision of the soul which had been turned downwards (κάτω; 519B). The knowledge of that which always is, as opposed to knowledge of that which comes into being and passes away, tends to draw the soul to truth (ἀλήθεια) and is productive of a philosophical attitude of mind, directing upwards (ἄνω) the faculties that are now wrongly turned downwards (κάτω; 527B). In Plato's dialogue, Socrates is unable to suppose that any other study would turn the soul's gaze upwards (ἄνω) than that which deals with being (τὸ ὄν) and the invisible (τὸ ἀόρατον). In his view, anyone who tries to learn about matters of the senses does not look up (ἄνω) but down (κάτω; 529B).

Similarly, in John's Gospel Jesus, in his dialogue with the Jews, having just asserted himself the world's true light (8:12), tells them: 'You are from below (ὑμεῖς ἐκ τῶν κάτω ἐστέ), I am from on high (ἐγὼ ἐκ τῶν ἄνω εἰμί)' (8:23). Jesus, as the one who has come from on high (ὁ ἄνωθεν ἐρχόμενος), is above all others, whereas he who is from the earth is earthly and uses earthly speech (3:31). This earthly, downward life, however, is turned upwards if one follows Jesus' imperative to be born from on high: Δεῖ ὑμᾶς γεννηθῆναι ἄνωθεν (3:7). And just as in Plato this upwards direction is concerned with learning (μαθήμα) concerning being (τὸ ὄν) and the invisible (τὸ ἀόρατον; 529B), in John, too, this upward life deals with instruction relating to the invisible God and the 'one who is' (ὁ ὤν) near, or from the side of God: Καὶ ἔσονται πάντες διδακτοὶ θεοῦ· πᾶς ὁ ἀκούσας παρὰ τοῦ πατρὸς καὶ μαθὼν ἔρχεται πρὸς ἐμέ. οὐχ ὅτι τὸν πατέρα ἑώρακέν τις εἰ μὴ ὁ ὢν παρὰ τοῦ θεοῦ (6:45b–46a; cf. 1:18 and 5:37–38).[59] Again, this shows that virtually the same didactic concern runs through John and book VII of Plato's *Republic*.

Of course, one could argue that Plato's *paideia* is more 'philosophical,' whereas John's didactics are of a more 'religious' nature. Yet, it may be anachronistic to play 'religion' and 'philosophy' off against one another. There seems to be a distinct language of conversion in Plato's allegory of the cave.[60] The ascent (ἐπάνοδος) from the subterranean

[59] Scholars agree that the designation of Christ as ὁ ὢν παρὰ τοῦ θεοῦ (6:46) or ὁ ὢν εἰς τὸν κόλπον τοῦ πατρὸς (1:18) is a clear allusion to the designation of God as ὁ ὤν, the 'One who is,' in *Exodus* 3:14: Ἐγώ εἰμι ὁ ὤν. In the Graeco-Roman period, the epithet ὁ ὤν was understood as a metaphysical designation for God. See M. Frede, 'Sein; Seiendes I. Antike,' in: *Historisches Wörterbuch der Philosophie*, vol. 9 (1995), 170–180, esp. I.2 Hellenismus; and Th. Kobusch, 'Sein; Seinendes II. Spätantike; Patristik,' 180–186. See also M. Burnyeat's paper in the forthcoming TBN volume on the Name.

[60] See Ph. Rousseau, 'Conversion,' in: *OCD* ³, 386–387; A.D. Nock, *Conversion: The*

cave to the sun follows the μεταστροφή, the turning from the shadows to the images that cast them and to the light (532B). This ascent to what really is (τοῦ ὄντος ἐπάνοδος), which is called 'true philosophy' (φιλοσοφία ἀληθής), is a περιαγωγή, a turning around of the soul, away from a nightly day towards the true, genuine day (521C; cf. 518C–E). It involves a process of being turned round (περιστρέφεσθαι) towards that which is unconcealed, true, and real (519B). It is a turning (μεταστροφή) from the world of generation to truth and essence (525C), an ἐπαναγωγή, a leading up of the soul (532C). This ascent (ἐπάνοδος) takes place along the road (ὁδός) which leads out of the cave (514A–B) and makes possible the soul's way up towards the intelligible region (ἡ εἰς τὸν νοητὸν τόπον τῆς ψυχῆς ἄνοδος; 517B). This ascension is what is meant in the parable by the upward ascension (ἡ ἄνω ἀνάβασις) and the sight of the things on high (θέα τῶν ἄνω; 517B).

This conversion language in Plato is very similar to the notion of being born from above (3:7: γεννηθῆναι ἄνωθεν) in John and his talk of Jesus as the way (ὁδός) which leads upwards to God's heavenly region (14:1–6) and along which God is seen (14:7–11). God has become visible inasmuch as Jesus has revealed himself as the world's true light (12:44–46). Conversion to Jesus, as John stresses in his description of the last instances of Jesus' public teaching, means converting to the unconcealed, true light (12:35–36; 12:46; cf. 8:12). Faith in Jesus (12:46) amounts to putting faith in the true light, with the consequence of becoming children of light: πιστεύετε εἰς τὸ φῶς, ἵνα υἱοὶ φωτὸς γένησθε (12:36).

Despite the general similarity of conversion language in Plato and John, an important difference arises in view of their evaluation of 'faith' (πίστις). In Plato's *Republic*, faith is but one step in the paideutic and dialectical process which advances through the stages of apprehension by means of images and shadows (εἰκασία), persuasion or faith (πίστις), understanding (διάνοια), and real knowledge (ἐπιστήμη) as opposed to mere opinion. In this dialectic process, which progresses by doing away with temporary hypotheses up to the first principle, the soul is led upwards (ἄνω) from the barbaric filth in which it is mired down (533C–534C).

Old and the New in Religion from Alexander the Great to Augustine of Hippo, London 1933, chap. 11: 'Conversion to Philosophy;' and A.D. Nock, 'Bekehrung,' in: *Reallexikon für Antike und Christentum*, vol. 2 (1954), 105–118.

Yet, there is remarkable agreement between Plato and John with regard to the function of dialectic. Dialectic can be defined as 'the science of conducting a philosophical dialogue (διαλέγεσθαι, "to converse") by exploring the consequences of premises asserted or conceded by an interlocutor,' and Plato's contribution to its development is his presentation of dialectic as 'co-operative investigation based on agreed premises,' which, also in his *Republic*, takes the form of oral debate.[61] In his *Republic*, dialectic is the supreme science. By dialectic (τῷ διαλέγεσθαι) one attempts through discourse of reason (διὰ τοῦ λόγου), and apart from all sensory perceptions, to find one's way to the very essence of each thing (532A).

In this sense, the longer dialogues of Jesus in John are dialectic, too. This has already been noted by C.H. Dodd, who emphasizes the contrast in form between the Johannine dialogues and those in the Synoptic gospels. Dodd assumes that the Johannine dialogues derive from the Hellenistic tradition, modelled on Plato's Socratic dialogues, of using 'dialogue as a vehicle for philosophical or religious teaching.'[62] Even though the interlocutor's role in John seems limited to misunderstanding, thus giving opportunities for the development of the dialogue, according to Dodd this also holds true for Plato's later dialogues, such as the *Timaeus*, in which 'the colloquy becomes little more than a device for introducing long monologues.'[63] A similar observation applies to book VII of Plato's *Republic*, in which Glaucon's role of interlocutor is very limited indeed.

The longer dialogues in John appear to be dialectic because they centre around particular sense-perceptible, physical, tangible actions or objects such as being born, water, and bread. In his dialogues Jesus uses words with multiple meanings, such as 'being born again/from above (ἄνωθεν)' (3:1–13), 'living water (ὕδωρ ζῶν)' (4:7–15), and 'bread

---

[61] D.N. Sedley, 'Dialectic,' in: *OCD* [3], 461.

[62] C.H. Dodd, 'The Dialogue Form in the Gospels,' *Bulletin of the John Rylands Library Manchester* 37 (1955) 54–67; quotation from p. 63. Cf. also R. Majercik, 'Dialogue,' in: *The Anchor Bible Dictionary* (1992), vol. 2, 185–188, esp. 187. Dodd is keen to stress that the fact that 'the evangelist has moulded his material into forms based upon current models of philosophical and religious teaching, instead of following the forms represented in the Synoptic Gospels' (p. 65) does not necessarily imply the unhistoricity of the material which John worked into his account. According to Dodd, 'it may well be that the still fluid tradition of the teaching of Jesus known to John included also material of which the Synoptic evangelists have taken no account, but which is of such a kind that it can be integrated with the Synoptic tradition' (pp. 66–67).

[63] Dodd, 'The Dialogue Form,' 62–64; quotation from p. 64.

from heaven' (6:26–59). The double entendre of these words occasions a further dialogue, in which the true, spiritual meaning of 'being born from above' (3:3, 7), 'living water' (4:10–11) and 'true bread' (6:32) is explored.[64] John seems to understand this kind of discourse as dialectic, because after Jesus' distinction between perishable food and true food (6:27, 55), his disciples react by saying: 'This way of reasoning (λόγος) is difficult' (6:60: Σκληρός ἐστιν ὁ λόγος οὗτος).[65] Just as Platonic dialectic aims at distinguishing perceptions of sense from the essence of each thing through discourse of reason (διὰ τοῦ λόγου; 532A), Jesus' longer dialogues in John are equally concerned with a dialectic discourse of reason (λόγος; 6:60) which is undertaken to establish the difference between what is physical and what is truly real (ἀληθινός; ἀληθής). This teaching of Jesus seems to constitute one more resemblance between John and the paideutic programme set forth in book VII of Plato's *Republic*.

According to John, the paideutic potential of Jesus' teaching seems also to be recognized by the Greeks themselves. It is during Jesus' last days in Jerusalem that some Greeks (Ἕλληνες), who happen to be in Jerusalem at the time, approach Philip, one of Jesus' pupils, and express their wish to meet Jesus (12:20–21).[66] Having received their

---

64 On double entendre in Jesus' discourses in John, see R. Kysar, 'John, the Gospel of,' in: *The Anchor Bible Dictionary* (1992), vol. 3, 916–917: section C3–4.

65 On the conscious use of λόγος in *John* 6:60, see the narrative analysis of John by M.W.G. Stibbe, *John's Gospel* (New Testament Readings), London & New York 1994, 24: 'The phrase translated "hard teaching" is *skleros logos*. It is a phrase which functions as a perfect title for Jesus who, in the prologue of John's gospel, is called God's *Logos*. In *John* 5–10, Jesus is truly the *Skleros Logos*, the Difficult Word.'

66 On the ethnic Greek identity of the 'Greeks' in John, see, e.g., C.R. Matthews, *Philip: Apostle and Evangelist: Configurations of a Tradition* (Supplements to Novum Testamentum 105), Leiden 2002, 114: 'It seems clear that the word Ἕλληνες must refer to gentiles, albeit proselytes, in view of the just voiced complaint of the Pharisees that the κόσμος is going after Jesus (12:19). Corroboration for this interpretation may also be found in Jesus' prediction concerning the drawing of all people to himself in 12:32 (also note 11:52). It is appropriate that this intriguing incident involves Philip and Andrew, the two disciples among the Twelve with Greek names.' See also J. Frey, 'Heiden—Griechen—Gotteskinder: Zu Gestalt und Funktion der Rede von den Heiden im 4. Evangelium,' in: R. Feldmeier and U. Heckel (eds), *Die Heiden: Juden, Christen und das Problem des Fremden* (WUNT I.70), Tübingen 1994, 228–268, esp. 250–251: 'Während (…) auf der Ebene der erzählten Geschichte in den Ἕλληνες Joh 12,20f. am ehesten Gottesfürchtige auf der Pilgerfahrt nach Jerusalem, in 7,35 hingegen heidnische Bewohner der griechischen Welt zu sehen sind, werden die Ἕλληνες in beiden Texten auf der allgemeingültigeren Ebene des johanneischen Symbolismus zu "Repräsentanten der griechischen Welt", ja zur Chiffre für die heidenchristlichen Adressaten des Evangeliums selbst.' Cf. also Josephus on Jesus' success among the Greeks (*Jew. Ant.* 18.63).

request (12:22), Jesus answers in a very indirect, non-concrete way (12:23a), talking about the prospect of bearing much fruit, the reward of following him, and the urgency of putting one's faith in the true light and becoming children of light (12:23b–36). This Greek perspective in John had already been introduced earlier in the Gospel, when Jews were said to ponder about the possibility of Jesus leaving Jerusalem for the 'Diaspora of the Greeks' with the purpose of teaching the Greeks (7:35–36): μὴ εἰς τὴν διασπορὰν τῶν Ἑλλήνων μέλλει πορεύεσθαι καὶ διδάσκειν τοὺς Ἕλληνας; This instruction of the Greeks, as John suggests, seems about to be realized at the very end of Jesus' public teaching. Even though Jesus' response to the request of the Greeks remains only indirect, John seems to highlight that the teachings of Jesus could satisfy Greek paideutic concerns, and that their contents have to do with his identity as the true light. This explicit focus on the 'Greeks' is absent from the Synoptic gospels,[67] and appears to be inseparable from John's interest in the light that enlightens all.

7. *The accessibility of the true light*

Despite all similarities, there is an important difference between John and Plato with regard to the light's accessibility. In principle, according to both authors, the true light enlightens all. Christ, in John's view, is the light which gives light to everyone (1:9: τὸ φῶς τὸ ἀληθινόν, ὃ φωτίζει πάντα ἄνθρωπον), and, according to Plato, those who receive Platonic *paideia* turn upwards the vision of their souls and fix their gaze on that which sheds light on all (540A: εἰς αὐτὸ ἀποβλέψαι τὸ πᾶσι φῶς παρέχον). Yet, in Plato's view the accessibility of the light is limited to the best natures, those capable of philosophy, who are forced to ascend from the cave into the light of day (519C–D; 520A) and are offered the fullest education possible (535A–540C).

This limited accessibility contrasts sharply with John's portrayal of the blind man and Lazarus as prototypes for each believer. All human beings, regardless of their intellectual potential, are invited to put their faith in the light and become children of light (12:35–36). In that sense, the parallelism between John and Plato is inverted, just as in the comparison between Jesus and Socrates. As Justin Martyr would put

[67] It is also John who tells us that the inscription on the cross was written in Hebrew, Latin, and Greek (19:20). The only reference in the Synoptic gospels to Greeks and Greek language is in Mark's story of Jesus' meeting with the Grecian woman in Syrophoenicia (*Mark* 7:26).

it: 'in Christ (...) not only philosophers and scholars believed, but also artisans and people entirely uneducated' (*Apologies* 2.10).[68]

Yet, notwithstanding the elitist nature of Platonic *paideia*, both John and Plato agree on the responsibilities of those who have seen the light. Neither John nor Plato has any Gnostic, world-denying inclinations. In Plato's *Republic*, the best natures who have been compelled to ascend towards the light and have received a better education than others are not allowed to linger outside the cave, but have eventually to take their turn to go down again to take charge of their former fellow prisoners (519C–D; 520C; 539E–540B). In Plato's imagery, the 'cave' into which the educated are sent down again (539E) symbolizes the cosmos. This is still implicit in Plato's *Republic*, but rendered explicit in later Platonist texts (see also Van den Berg, this volume, §1).[69]

In John, this world-affirming attitude is mirrored in Jesus' final prayer on behalf of his pupils immediately before his capture, trial, and death. In this prayer, Jesus does not ask God to take his pupils away from this cosmos, but to consecrate them by the truth (ἀλήθεια) now they are being sent into the cosmos. 'As you sent me into the cosmos,' Jesus tells God, 'I have sent them into the cosmos:' καθὼς ἐμὲ ἀπέστειλας εἰς τὸν κόσμον, κἀγὼ ἀπέστειλα αὐτοὺς εἰς τὸν κόσμον (17:15–18). Just as Socrates orders the educated to go down (520C: καταβατέον οὖν ἐν μέρει ἑκάστῳ εἰς τὴν τῶν ἄλλων συνοίκησιν), in a similar way Jesus sends his pupils into the world. Both Plato's and John's enlightenment do not aim at a retreat from this world, but at shedding light on the proper hierarchy of physical and non-physical, spiritual levels within this world. This holds true for John, too, as, after all, this visible cosmos has been created through the true light (1:10).

---

[68] For this non-elitist self-understanding of Christianity as opposed to the elitism of Greek philosophy, despite similarity in content, see also Origen, *Against Celsus* 7.42–43. Both Justin and Origen demonstrate the difficulty of Greek philosophy by referring to Plato, *Timaeus* 28C: 'Now to discover the Maker and Father of this universe were a task indeed; and having discovered Him, to declare Him unto all men were a thing impossible' (transl. R.G. Bury). Cf. Justin, *Apologies* 2.10 and Origen, *Against Celsus* 7.43. On this use of *Timaeus* 28C, see A.D. Nock, 'The Exegesis of Timaeus 28C,' *Vigiliae Christianae* 16 (1962) 79–86. See also Van den Berg, this volume, §3.

[69] See, e.g., Numenius, frg. 60; Plotinus, *Enneads* 4.8.3; and Porphyry, *The Cave of the Nymphs in the Odyssey* 5.1, 6.11–20, and 8.12.

3. *Conclusion and final observations*

John's identification of Christ as the world's true light in the Prologue to his Gospel is part of his conscious modelling of the Prologue on the opening of *Genesis*. Reading about the invisibility of the earth in the Septuagint translation of *Genesis* 1:2 (ἡ δὲ γῆ ἦν ἀόρατος), it seems plausible that John—like Philo and Clement before and after him—took the invisibility of this earth to allude to the non-visible, noetic paradigm which was subsequently implemented in the visible world at its creation. For that reason, John also took the reference to the light in *Genesis* 1:3 as a reference to the invisible, true, real light which preceded the creation of the world's physical light. The concept and terminology of true, real, noetic light was at home in Platonist thought and derives ultimately from *Phaedo* 109E. John introduces this Platonic notion of the true light in his Prologue (1:9) and links it up with Plato's further elaboration on this light as the light which enlightens all in his *Republic* (540A).

As this connection between John and Plato seems to remain unnoticed in modern scholarly literature, it seems relevant to point out that Church fathers such as Origen and Augustine, who were still imbued with the ideas and arguments of classical philosophy, had no difficulty in recognizing it. According to Origen, the Platonic idea that 'a light suddenly arrived in the soul as though kindled by a leaping spark' is contained in John's assertion that Christ, the Logos, is 'the light of men,' which—Origen adds—is 'the true light that lightens every man coming into the true and intelligible world' (*Against Celsus* 6.5).

The same link between the Platonic notion of the true, noetic light and John is present in Origen's polemic against the worship of the heavenly bodies. In this polemic, Origen stresses that it is unreasonable that human beings

> should have been amazed at the visible light of the sun, moon and stars (τὸ αἰσθητὸν ἡλίου καὶ σελήνης καὶ ἄστρων φῶς), to such an extent that because of their visible light they should somehow regard themselves as beneath them and worship them. For they (human beings, that is) possess a great intellectual light of knowledge (τηλικοῦτον νοητὸν γνώσεως φῶς) and a 'true light' (φῶς ἀληθινόν; *John* 1:9), and a 'light of the world' (φῶς τοῦ κόσμου; *John* 8:12; 9:5; cf. 11:9), and a 'light of men' (φῶς τῶν ἀνθρώπων; *John* 1:4). If they ought to worship them, they ought not to do so because of the visible light which amazes the masses (οὐ διὰ τὸ θαυμαζόμενον ὑπὸ τῶν πολλῶν αἰσθητὸν φῶς), but because of the intellectual and true light (ἀλλὰ διὰ τὸ νοητὸν καὶ ἀληθινόν), supposing

> that the stars in heaven are also rational and good beings (...). However, not even their intellectual light ought to be worshipped by anyone who sees and understands the true light (τὸ ἀληθινὸν φῶς) (...). Those who have realized how 'God is light' (*1 John* 1:5), and who have comprehended how the Son of God is 'the true light, which lightens every man coming into the world' (*John* 1:9), and have also understood what he meant when he said 'I am the light of the world' (*John* 8:12), would not reasonably worship the light in the sun, moon and stars which is like a dim spark compared with God who is light of the true light
> (*Against Celsus* 5.10–11; transl. H. Chadwick).[70]

In a similar way, Augustine criticizes the inconsistency of those Platonic philosophers who suppose that many gods are to be worshipped. According to Augustine, this is inconsistent because the Platonists themselves agree that 'the soul of man' and the 'immortal and blessed dwellers in heaven' derive their blessedness from the same source,

> from a certain intelligible light cast upon them, which is their God, and which is different from themselves, and which illuminates them so that they are enlightened, and may by their participation in it exist in a state of perfect blessedness (*The City of God* 10.1–2; quotation from 10.2).[71]

To demonstrate this basic agreement between him and the Platonists, Augustine points to Plotinus who, in his explanation of Plato, asserts that not even

> the soul of the cosmos derives its blessedness from any other source than does our own soul: that is, from the light which is different from it, which created it, and by whose intelligible illumination the soul is intelligibly enlightened (*The City of God* 10.2).[72]

All beings receive their blessed life and 'the light by which the truth is understood' from the same source. 'This,' as Augustine explicitly says, 'is in harmony with the Gospel,' and he goes on to quote the passage from John's Prologue on 'the true light which lighteth every man that cometh into the world' (*John* 1:9). From this Augustine draws

---

[70] Cf. also Origen, *Commentary on John* 1.159–161, 164, and 167. On this issue in Origen, see also A. Scott, *Origen and the Life of the Stars: A History of an Idea* (Oxford Early Christian Studies), Oxford 1991, 132–133; and Wallraff, *Christus Verus Sol*, 42, 54–55, 57, 64.

[71] Transl. R.W. Dyson, *Augustine: The City of God Against the Pagans* (Cambridge Texts in the History of Political Thought), Cambridge 1998.

[72] Cf. Plotinus, *Enneads* 4.7.10, 6.9.4, and 6.9.9. For the continuing influence of this idea in later philosophy, see N. Jolley, *The Light of the Soul: Theories of Ideas in Leibniz, Malebranche, and Descartes*, Oxford 1990.

the conclusion that 'the rational or intellectual soul (…) cannot be its own light, but shines by its participation in another and true light' (*The City of God* 10.2).

Both Origen and Augustine explicitly link the Platonic notion of the 'true light' with John. It is probable, however, that this link is not due to Origen's and Augustine's Greek-philosophical interpretation of John, but to the fact that John himself drew on Platonic philosophy. The high esteem in which Plato was held by contemporary Judaism, even in Galilee as the case of Justus of Tiberias demonstrates, makes this far from unlikely. John seems to have had access to Greek *paideia*. Like Philo, John seems to have taken note of important themes and issues in the allegory of the cave in Plato's *Republic*. This is suggested by the way in which John elaborates the concept of the true light in the body of his Gospel, after he has introduced it in the Gospel's Prologue in which he interprets the light mentioned in *Genesis* 1:3 as the true, archetypal light of the invisible, paradigmatic creation. The dynamics of the light's effulgence and people's receptiveness to the 'true light' and to 'truth' constitute a running theme within John's Gospel.

CREATION 'OUT OF' AND 'THROUGH' WATER IN 2 PETER 3:5

Edward Adams

1. *The creation of the world in Second Peter*

In 2 Pet 3:5, the author of this deeply fascinating letter makes reference to the creation of the world.[1] He[2] does so in a highly striking and unusual way, speaking of God's formation (συνεστῶσα)[3] of the heavens and the earth by his word, 'out of water and through water', ἐξ ὕδατος καὶ δι᾽ ὕδατος.

The statement occurs in the context of a debate in 2 Peter 3 between the author and his opponents whom he calls 'scoffers' (cf. 3:3).[4] The author cites their claims in 3:4 and responds to them in 3:5–13.[5] The controversy is not at all about creation, but the parousia and the future of the cosmos. The author's adversaries mock the promise of the Lord's coming on account of its non-fulfilment: the long period of time that has passed since the promise was initially made shows that

---

[1] This is a revised version of the paper given at the 'Themes in Biblical Narrative Conference' in Groningen, June 13th-14th, 2003. I would like to thank Dr George van Kooten for inviting me to participate in this conference and to contribute to this volume. I should also thank conference participants for their responses to my paper which have helped me to shape the presentation of it as it now appears.

[2] Second Peter is generally regarded by scholars as a pseudonymous work of the late first or early second century. The letter was written to counteract the influence of those whom the author views as false teachers (2:1). The polemic which forms the central part of the epistle in chapter 2 is one of the fiercest pieces of invective in the New Testament.

[3] The participle συνεστῶσα belongs grammatically with γῆ, but since the writer obviously believes that the heavens, as well as the earth, were created by God, the participle ought to be taken with both οὐρανοί and γῆ. E.g., R.J. Bauckham, *Jude, 2 Peter* (Word Biblical Commentary 50), Waco 1983, 296.

[4] The scoffers (the false teachers of 2:1) are presented in 3:3 as a phenomenon of 'the last days', but it is clear from what follows (and from the polemic of chapter 2) that that they are the writer's contemporaries and that the debate depicted is a current one.

[5] Whether or not the writer is reporting the actual words, the *ipsissima verba*, of his opponents, it is highly probable that the citation is an accurate representation of the claims they were making, giving us the *ipsissima vox* (the very voice) of the opponents.

the expectation is a forlorn one. In addition, they assert, 'All things remain as they were from the beginning of creation' (3:4).[6] According to the opponents, the world has survived since creation without major change. The visible stability of the cosmos and its constancy over a very long period of time show that it is an everlasting structure.[7] Evidently, they understood the parousia to involve the destruction of the universe.[8] For them, the cosmos is immutable and imperishable. It is this point with which the author deals first, in 3:5–7. In 3:8–9, he takes up the question of God's apparent slowness in fulfilling his promise. Then in 3:10, he describes the fiery end of the world, before focusing in 3:11–13, on the moral implications of the coming dissolution.

The main argument in 3:5–7 is not difficult to discern. It is not the case, the writer maintains, that the world has continued from creation without significant alteration. It was destroyed long ago at the flood (3:6). If the cosmos was devastated once before, it can be destroyed again; it is not by nature everlasting.[9] At the appointed time, it will be dissolved by fire (3:7). The idea that the flood is a typological precursor of the final judgement (which is found already in Isaiah 24:18) seems to be implicit. The reference to creation in 3:5 does not appear to contribute much to this line of argument; the case rests on a correlation between the flood and the coming judgement. However, there is a further level of argumentation in 3:5–7 to which the mention of creation is integral. Creation, flood and the final judgement are linked, the author indicates, by the divine word.

---

[6] 2 Peter 3:4: πάντα οὕτως διαμένει ἀπ' ἀρχῆς κτίσεως.

[7] So, e.g., C. Bigg, *A Critical and Exegetical Commentary on the Epistles of St. Peter and St. Jude* (International Critical Commentary), Edinburgh 1901, 292. Bauckham, *Jude, 2 Peter*, 293 and other recent commentators think that the words πάντα οὕτως διαμένει ἀπ' ἀρχῆς κτίσεως amount to a denial of divine intervention rather than an assertion of cosmic indestructibility. For criticisms of this line of interpretation, see E. Adams, 'Where is the Promise of His Coming? The Complaint of the Scoffers in 2 Peter 3:4', *New Testament Studies* (forthcoming).

[8] Global or cosmic upheaval is frequently associated with the eschatological coming of God in Jewish apocalyptic texts: *1 Enoch* 1:2–9; 102:1–3; *Testament of Moses* 10:3–7; *Biblical Antiquities* 19:13; cf. Micah 1:2–4; Nahum 1:3–5; Habakkuk 3:3–15. The language of cosmic catastrophe is applied to the coming of the Son of Man in Mark 13:24–25 + par.

[9] As Bauckham observes, the writer conceives of the Noahic flood as a total cosmic disaster (Bauckham, *Jude, 2 Peter*, 299). Cf. *1 Enoch* 83:3.

The argument advances in three steps.[10] First, it was by the word of God (τῷ τοῦ θεοῦ λόγῳ) that heavens and earth were made (3:5); second, the divine word, along with water (δἰ ὧν), caused the world to be deluged (3:6) in the time of Noah;[11] third, by the same word (τῷ αὐτῷ λόγῳ), God has decreed that the world as it now is will end in a cosmic blaze on the day of judgement (3:7). The author thus argues for world-destruction *from* belief in God's creation of the world—a belief also held by his opponents (cf. 3:4). Since the world has been *created* by God's command, it can also be *de-created* by it. God destroyed it once before at the flood; he will do so again at the coming judgement.

There is yet another level of argumentation in 3:5–7, and it is to this argument that the reference in 3:5 to creation out of and through water (rather than creation by divine word) contributes. Again, the line of thought appears to develop in three steps, the first two forming the basis for the third: first, the world was formed *out of and through water* (ἐξ ὕδατος καὶ δἰ ὕδατος, 3:5); second, at the flood, the world was destroyed *by water* (ὕδατι, 3:6); third, the world is now reserved *for fire* (πυρί, 3:7). However, the logic here is not so easy to follow. How does the third point follow from points one and two? While the wording ἐξ ὕδατος καὶ δἰ ὕδατος in 3:5 paves the way for ὕδατι in 3:6 and sets up an obvious linkage between creation and flood, it is not clear how these references to ὕδωρ prepare for the mention of πῦρ in 3:7.

The words ἐξ ὕδατος καὶ δἰ ὕδατος in 3:5, are normally explained by commentators against the background of Genesis 1. The influence of the Genesis creation narrative on this verse is evident from the mention of 'heavens and earth' (echoing Gen 1:1), the reference to water (echoing Gen 1:2, 6, 7, etc.) and the stress on God's word as the instrument of creation (Gen 1:3, 6, 9, etc.).

The first phrase, ἐξ ὕδατος, is taken by most modern commentators as an allusion to the emergence of the heavens and earth from the pre-existent waters of chaos as described in Genesis 1. According to the Genesis narrative, the primeval waters (Gen 1:2) were divided into two, thereby forming a vault—the sky—with water above and below it (Gen 1:6–8). The lower waters were then gathered together, revealing

---

[10] D.G. Horrell, *The Epistles of Peter and Jude*, Peterborough 1998, 176–177 rightly discerns a three-step argumentation in 2 Pet 3:5–7, but does not distinguish the different levels on which the author argues.

[11] The phrase δἰ ὧν in 3:6 most probably refers to both water and the word of God: cf. Bauckham, *Jude, 2 Peter*, 298.

the dry land (Gen 1:9–10), which God called the earth. This notion of a primeval sea is a generic feature of ancient near-eastern mythological cosmogony. The thought expressed by the writer's wording, if taken literally (as we will see below), goes beyond the description in Genesis, but the phrase ἐξ ὕδατος is at least, as Kelly puts it, 'an understandable gloss' on it.[12] The complementary phrase, δι' ὕδατος, on the other hand, is very difficult to explain purely in terms of Genesis 1. None of the proposed attempts to do so fully satisfy.

An older explanation is that the writer is using the preposition διά with the genitive in a local sense—'in the midst of' or 'between'.[13] However, the use of διά + genitive with this meaning, though not unprecedented, would be highly unusual.[14] Significantly, the LXX of Gen 1:6 uses the constructions ἐν μέσῳ and ἀνὰ μέσον to relate the creation of the sky 'in the midst of' and 'between' the waters.[15]

Another suggestion is that δι' ὕδατος has an instrumental sense and refers to the means by which the earth was sustained, i.e., rain (cf. Gen 2:5).[16] But the reference in 2 Pet 3:5 is clearly to the *formation* of the world, not to the sustenance of life upon it.[17]

The standard approach among commentators is to take δι' ὕδατος instrumentally, 'by means of water', and to see it as a rather nebulous reference to God's separation of the waters. Thus Bauckham writes, 'the writer means that water was, in a loose sense, the instrument of creation, since it was by separating and gathering the waters that God created the world.'[18] But this is a very strained attempt to make the language fit Genesis 1.

The difficulty commentators have in making sense of δι' ὕδατος against the background of the Genesis creation narrative raises the

[12] J.N.D. Kelly, *A Commentary on the Epistles of Peter and Jude* (Black's New Testament Commentaries), London 1969, 358.

[13] E.g., J.B. Mayor, *The Epistle of St. Jude and the Second Epistle of St. Peter*, London 1907, 151; J. Chaine, 'Cosmogonie aquatique et conflagration finale d'après la *secundri Petri*', *Revue biblique* 46 (1937), 207–216, esp. 210 note 3.

[14] Cf. Kelly, *Peter and Jude*, 358–359.

[15] Gen 1:6 LXX: Γενηθήτω στερέωμα ἐν μέσῳ τοῦ ὕδατος καὶ ἔστω διαχωρίζον ἀνὰ μέσον ὕδατος καὶ ὕδατος. καὶ ἐγένετο οὕτως.

[16] E.g. Bigg, *St. Peter and St. Jude*, 293; M. Green, *The Second Epistle General of Peter and the General Epistle of Jude* (Tyndale New Testament Commentaries), Leicester 1987 (Revised ed.), 141.

[17] The verb συνιστάναι can mean 'subsist', as in Col 1:17, but here the sense is plainly 'put together', 'form'.

[18] Bauckham, *Jude, 2 Peter*, 297.

question of whether the author in his coupling of the phrases ἐξ ὕδατος and δἰ ὕδατος is dependent on another creation tradition, alongside the biblical one.

Tord Fornberg, in his landmark study of Second Peter in its Hellenistic context, points out that the verb συνιστάναι is a technical creation term in Greek and Hellenistic authors.[19] Used in connection with the preposition ἐκ, it normally indicates the material out of which the cosmos was made.[20] With the construction ἐξ ὕδατος ... συνεστῶσα, therefore, the author seems to be saying that water was the very stuff out of which the cosmos was created. Such a notion, of course, exceeds the teaching of Genesis 1. Bauckham thinks that the scientific sense of the language should not be pressed; the author means to convey no more than the mythological concept found in Genesis 1, that the world emerged from a primeval sea.[21] But the choice of words is surely significant. The author of Second Peter is culturally sophisticated[22] and, elsewhere in his letter (especially 1:3–11) shows a willingness to employ Hellenistic religious and philosophical language to express his theological ideas. It is highly likely that he is well aware of the scientific implications of his wording and that he is deliberately evoking them.

In 2 Pet 3:5, therefore, as well as calling to mind Genesis 1, the writer appears to be alluding to a particular Greek or Hellenistic cosmological tradition, one in which water is specifically identified as the substance out of which the world was made. That tradition, one would suppose, would give clarity to the enigmatic δἰ ὕδατος.

Thales, the Milesian natural philosopher of the sixth century BCE, is credited with the view that water is the ἀρχή, the beginning of all things, the element out of which everything else has emerged.[23] Kelly thinks that the author of Second Peter is interpreting Genesis 1 in the

---

[19] T. Fornberg, *An Early Church in a Pluralistic Society: A Study of 2 Peter* (Coniectanea Biblica, New Testament Series 9), Lund 1977, 67.

[20] See references in Fornberg, *An Early Church*, 67 note 12.

[21] Bauckham, *Jude, 2 Peter*, 297.

[22] He writes in a style of Greek known as 'Asiatic Greek' which was still fashionable at the time he was writing: see B. Riecke, *The Epistles of James, Peter, and Jude: Introduction, Translation and Notes* (Anchor Bible), New York 1964, 146–147.

[23] Aristotle, *Metaphysica* A3, 983b6ff. (= Diels-Kranz 11 A12). On Thales' cosmology, see G.S. Kirk, J.E. Raven and M. Schofield, *The Presocratic Philosophers: A Critical History with a Selection of Texts*, Cambridge 1983², 88–94. Thales' view was probably a scientific rationalisation of the notion found in Homer that 'ocean' was the begetter of all things, including the gods (*Iliad* 14.201, 246). This in turn was probably a development of the Ancient Near Eastern mythology of the earth arising from the primeval waters.

light of Thales' doctrine.[24] However, it is not obvious why the writer would set so much store by Thales' opinion. Thales' views on cosmic origins were certainly treated with respect but hardly represented the cutting edge of scientific thinking in the late first or early second century CE. Also, from what we know about Thales' teaching, there is nothing in it that would illuminate the phrase δἰ ὕδατος. A modified version of Thales' notion, though, was adopted by the early Stoics and formed part of their cosmogony. Stoicism was still, of course, a highly influential philosophical system when Second Peter was written; its main teachings would have been quite accessible to our author. In 1:5–7, the author shows an acquaintance with Stoic ethical terminology.[25] It is my contention in this essay that the writer is drawing on the characteristically Stoic view of world-formation in 2 Pet 3:5. A Stoic background, it seems to me, explains both the phrase ἐξ ὕδατος and the otherwise obscure δἰ ὕδατος. It also illuminates the three-stage argument in 3:5–7 involving ἐξ ὕδατος καὶ δἰ ὕδατος, ὕδατι and πυρί.

In what now follows, I first give an account of Stoic cosmogony, focusing particularly on the role of water in the scheme (§2). I then try to show that the creation-formulation of 2 Pet 3:5 is consistent with this tradition (§3). Next, I look briefly at the writer's utilisation of Stoic cosmology in 3:6–7 (§4). Then, I attempt to clarify the author's line of reasoning from ἐξ ὕδατος καὶ δἰ ὕδατος to ὕδατι to πυρί, in the light of his dependence on Stoic physics (§5). The writer's deployment of such an argument, I suggest, is explicable in the context of a debate about the future of the cosmos, which is partly philosophical in nature (§6).

2. *Stoic cosmogony*

The Stoic view of the world's origins, as David Hahm has shown in his milestone study of Stoic cosmology, has both a physical and a biological dimension.[26] The world's emergence could be explained in terms of pure physics. It could also be described in biological terms, on the analogy of human conception and birth. The two accounts represent the same underlying process but from different perspectives. The world's appearance out of water is a prominent feature of both models.

---

[24] Kelly, *Peter and Jude*, 358.

[25] See further Bauckham, *Jude, 2 Peter*, 174–176.

[26] D.E. Hahm, *The Origins of Stoic Cosmology*, Columbus, Ohio 1977, 57–90.

The Stoics held to a cyclic view of cosmic history: the ordered world arises, continues for a period, ends in a massive conflagration and is recreated anew, the sequence repeating itself endlessly.[27] In later Stoic thought, we find the belief that the world ends in alternate catastrophes by flood and fire (see further below).

The primary state of the universe, according to the Stoics, is one of utter fire—the condition of all things at the conflagration. During the process of world-generation, the fire goes through a series of material transformations until eventually the settled cosmos as we know it comes into being. On the strictly physical account of the genesis of the cosmos, the world reaches its present form via the route of elemental change. Chrysippus, in a quotation from his first volume on *Physics* preserved by Plutarch, describes the process as follows:

> The transformation of fire is like this: by way of air it turns into water (δὶ ἀέρος εἰς ὕδωρ τρέπεται); and from this, as earth is precipitated, air evaporates; and as the air is subtilized, ether is diffused round about, and the stars along with the sun are kindled from the sea.[28]

A report given by Diogenes Laertius, in his discussion of Zeno and Stoicism, corroborates the scheme.

> The world, they hold, comes into being when its substance (ἡ οὐσία) has first been converted from fire through air into moisture (δὶ ἀέρος εἰς ὑγρότητα) and then the coarser part of the moisture has condensed as earth, while that whose particles are fine has been turned into air, and this process of rarefaction goes on increasing till it generates fire. Thereupon out of these elements animals and plants and all other natural kinds are formed by their mixture.[29]

The first change is the transition from fire to air. The original fiery matter of the cosmos cools down and becomes airy. The air then changes into water. At this juncture, all that is visible is liquid. As Hahm states, if an observer existed, 'he would *see* only water.'[30] Out of this watery stuff, the four elements of which the cosmos is composed are produced. The weightier part of water condenses and solidifies to make earth. The lighter part evaporates producing the element air. The air

---

[27] For representative primary sources and brief commentary, see A.A. Long and D.N. Sedley, *The Hellenistic Philosophers: Translations of the Principal Sources with Philosophical Commentary*, Cambridge 1987, 274–279.

[28] Plutarch, *On Stoic Self-Contradictions* 1053A.

[29] Diogenes Laertius, *Lives of Eminent Philosophers* 7.142.

[30] Hahm, *Origins*, 83 note 2.

then thins out generating the element fire. What is left continues as water. The four elements combine in many diverse ways to make the variety of things and life-forms on earth.

In developing this scheme, the Stoics drew on previous cosmological theories, but combined them in a distinctive way. They took from the Milesians of the sixth century BCE, the basic idea that the world developed from a single undifferentiated substance.[31] They drew specifically on Thales' teaching for their view that all things derive most directly from water (see also Bremmer, this volume, §1.1). From Anaximenes, they borrowed the notion of the growth of the elements from a single root through the twin processes of condensation and rarefaction.[32] The belief that fire is the archetypal form of matter they adapted from Heraclitus.[33]

The physical account of cosmic origins is mechanistic. One element changes into another apparently through its own natural propensity. There is no explicit reference to any force acting upon the substance to bring about the metamorphosis. On the biological and teleological explanation, however, there is an energizing force driving the changes and ordering things. The world's origins is construed in terms of an active principle working on a passive one. The active principle is God or Zeus; the passive principle is unshaped matter. God acts on the matter to bring to birth the physical cosmos. Diogenes explains the model.

> In the beginning he [God] was by himself; he transformed the whole of substance through air into water, and just as in animal generation the seed has a moist vehicle, so in cosmic moisture God, who is the seminal reason of the universe, remains behind in the moisture as such an agent, adapting matter to himself with a view to the next stage of creation. Thereupon he created first of all the four elements, fire, water, air, earth.[34]

---

[31] According Thales, the ἀρχή was water. Anaximander took it to be the 'infinite' (Diels-Kranz 12 A9, 10, 11, 14) and Anaximenes contended it was air (Diels-Kranz 13 A5, 7, 8). On Anaximander, see Kirk, Raven and Schofield, *The Presocratic Philosophers*, 105–117; On Anaximenes, see Kirk, Raven and Schofield, *The Presocratic Philosophers*, 144–148.

[32] Diels-Kranz 13 A6; cf. Kirk, Raven and Schofield, *The Presocratic Philosophers*, 148–150.

[33] It is unlikely, however, that Heraclitus viewed fire as an originating principle as Thales had viewed water and Anaximenes air: see Kirk, Raven and Schofield, *The Presocratic Philosophers*, 198.

[34] Diogenes Laertius, *Lives of Eminent Philosophers* 7.136.

In this account, God actively transforms the initial sole substance through the stages of world-formation into the variegated cosmos. God is the creative power *within* matter, working upon it, making it adaptable to his purposes, taking it from one stage of creation to another. Now from a Stoic point of view God is not transcendent over matter, but is entirely co-extensive with it. He (or it) is the rational principle (λόγος) in matter, theoretically distinguishable from it, but in reality absolutely inseparable from it.[35]

As Hahm points out, Diogenes' description 'bristles with biological terms.'[36] The origin of the universe is likened to the reproduction of living creatures, especially humans. The water into which the generative fire turns is compared to male seminal fluid. According to Stoic biology, male semen is made up of water and an active, hot generative force called 'seed' (σπέρμα).[37] The fluid is the vehicle for the seed, which is the actual reproductive component. So it is with the conception of the cosmos. The moisture which is formed when the cosmic fire abates consists of wet matter and seed. The liquid is the medium for the reproductive force. God is the seed itself, the formative power in the primordial water.

In a passage in the thirty-sixth discourse of Dio Chrysostom, the biological analogy is taken further. Here, the cosmogonal scheme is elucidated in the form of an allegorical myth, the so-called 'holy wedding' of Zeus and Hera.[38]

> But recalling Aphrodite and the process of generation, it [Zeus] tamed and relaxed itself and, quenching much of its light, it turned into fiery air of gentle warmth, and uniting with Hera and enjoying the most perfect wedlock, in sweet repose it emitted anew the full supply of seed for the universe ... And having made fluid all his essence, one seed for the entire world (ἓν σπέρμα τοῦ παντός), he himself moving around in it like a spirit (πνεῦμα) that moulds and fashions in generation, then indeed most closely resembling the composition of the other creatures, inasmuch as he might with reason be said to consist of soul and body, he now with ease moulds and fashions all the rest, pouring about him his essence smooth and soft and easily yielding in every part.[39]

---

35 See further M. Lapidge, 'Stoic Cosmology', in: J.M. Rist (ed.), *The Stoics*, Berkeley 1978, 160–185, esp. 163–164.

36 Hahm, *Origins*, 60.

37 Hahm, *Origins*, 68.

38 Cf. Homer, *Iliad* 14.294–296.

39 Dio Chrysostom, *Discourses* 36.56–57. In this passage, the 'seed' is not the active, fiery principle *in* the cosmic semen, but the semen itself.

Zeus and Hera engage in sexual intercourse, and Zeus emits the seminal fluid which is the seed of the universe. Zeus himself passes through Hera into the semen and becomes the creative energy within it. From inside, he shapes the wet matter and fashions it into the ordered cosmos.

It emerges from Dio's allegory that when the fire changes into water, the liquid mass is considered to consist of soul and body. Clearly, the divine principle is the soul, and the wet matter the body. During the fiery phase of the cosmic history, all that exists is soul. It is only when the fire turns into water that the totality becomes corporeal. This is confirmed by a statement made by Chrysippus in his first book on *Providence* preserved by Plutarch. When the fire is transformed into liquid, he states, 'it has in a way changed into body and soul so as to be a composite of these' (Plutarch, *On Stoic Self-Contradictions* 1053B).

The role of water in the Stoic explanation of the world's origins may now be clarified. Water is not the primary element, the absolute material source of all that exists; that distinction belongs to fire. The watery period of cosmic growth is *a transitional phase* in the emergence of the world. It is not the original state of all things, but one of the material transformations undergone by 'the whole' as it changes from pure fire to the physical world with which we are familiar. Water is, however, *the immediate substance* out of which the world is formed. When the fire becomes water, by way of air, it becomes pliable material, stuff with which the divine craftsman can work. God moulds it, adapts it, deriving from it other forms of matter, and fashions from it the ordered world of common experience.

In Stoic cosmogonal thought, the change to water is properly *the beginning of our world*. It is the moment of the world's conception: this is clear from the comparison of the cosmic water to male reproductive fluid and is vividly expressed in Dio's 'orgasmic' imagery. Although, within the Stoic scheme, the transition to air precedes that to water, it is connected more with the abatement of the conflagration than with the emergence of a new world. Little is made of it in the biological explanation of how the world began. In Dio's allegory, Hera or air, 'has the function of inducing the emission of seed, but contributes nothing to the offspring.'[40] That the shift to water constitutes the point at which a new cosmos starts to materialize is confirmed by Seneca; he writes,

[40] Hahm, *Origins*, 62.

> I will add, as Thales says, 'Water is the most powerful element.' He thinks it was the first element, and all things arose from it. We Stoics are also of this opinion, or close to it. For we say that it is fire which takes possession of the universe and changes all things into itself; it becomes feeble, fades, and sinks, and when fire is extinguished nothing is left in nature except moisture, in which lies the hope of the universe to come. Thus, fire is the end of the world, moisture the beginning.[41]

3. *2 Pet 3:5 as an allusion to Stoic cosmogony*

The double prepositional construction ἐξ ὕδατος and δἰ ὕδατος makes very good sense against the background of Stoic cosmogony. On the basis of the Stoic account of cosmic origins, it would be quite correct to say that the cosmos was formed 'out of' water, since water, though not the archetypal element, was nevertheless the immediate substance out of which the cosmos was made, the malleable, corporeal stuff which the divine craftsman shaped and adapted into an ordered world. It would be equally correct to say that the heavens and the earth were formed 'through' water, since water was not the original state of things but one of the material alterations experienced by the universe on its way to becoming a fully formed structure. On this understanding, διά + genitive does not have the instrumental sense 'by means of', which most commentators ascribe to it, but has the sense of 'through the medium of', denoting a temporary material state.[42] The phrase δἰ ὕδατος, admittedly, is not attested in the sources cited above with reference to the transformation to water, so there is no exact verbal parallel to it in the Stoic tradition. Yet, the words δἰ ἀέρος in the two extracts from Diogenes above serve as a close comparison, indicating as they do a transitional step in the process that leads from fire to life-sustaining cosmos.

With his words ἐξ ὕδατος καὶ δἰ ὕδατος, therefore, I suggest that the author is alluding to the Stoic cosmogonal scheme. From the creation traditions available to him, it would not be easy to find a more exact conceptual fit to his phrasing.

---

[41] Seneca, *Natural Questions* 3.13.

[42] That this would give διά + genitive a different sense from its use in the very next verse (δἰ ὧν 'by means of which') is not a strong objection. In 1:4, the writer uses ἐν + dative twice in back to back phrases to express two different meanings: ἐν τῷ κόσμῳ ἐν ἐπιθυμίᾳ. In the first phrase ἐν is locative, 'in the world'; in the second, the preposition is causal, 'because of desire'.

In making this allusion, the author is not giving his unqualified approval to the Stoic theory of world-formation. He could hardly have subscribed to the biological version of the theory, with its view of God as utterly conjoined with matter, as an impersonal, energizing power operating within, and as the active male component in the generative process. He is, though, I would suggest, committed to those aspects of Stoic cosmogonal teaching connoted by the twin prepositional constructions: first, that water was the direct elemental substance out of which everything else was made; second, that water was not the original state of this substance, but a material transformation of a more fundamental element. Since the writer explicitly mentions fire in 3:7, it is reasonable to suppose that, in line with Stoic doctrine, he took that basic element to be fire.

In 2 Pet 3:5, the author is thus attempting, in a limited way, to integrate Genesis 1 with Stoic physics. He takes the correlation no further than the equation of the watery waste of Gen 1:2 with the water out of which, according to Stoic cosmogony, the world arose. As noted above, in Stoic thought, the watery phase of cosmic existence was properly the 'beginning' of the world as a corporeal reality. This, perhaps, facilitated a connection with the 'beginning' in Gen 1:1. The most striking feature of the writer's Stoic re-interpretation of the Genesis creation narrative is the implication that there was a situation before that described in Gen 1:2: before there was a watery chaos, there was a fiery mass.

4. *2 Pet 3:5–7 as reflecting a Stoic scheme of cosmic history*

It is not only at 3:5 that the writer attempts to correlate biblical tradition with Stoic cosmology. In 3:7, he links the biblical notion of a coming day of judgement (cf. 'day of the Lord', 3:10; 'day of God', 3:12) with the Stoic expectation of a cosmic conflagration.[43] In Old Testament prophecy, fire is often associated with divine judgement,[44] but the

[43] On the Stoic concept of *ekpurosis*, or cosmic conflagration, see A.A. Long, 'The Stoics on World-Conflagration', *The Southern Journal of Philosophy* 23 (1985), 13–37. See also the earlier, excellent study by J. Mansfeld, 'Providence and the Destruction of the Universe in Early Stoic Thought', in: M.J. Vermaseren (ed.), *Studies in Hellenistic Religion*, Leiden 1979, 129–188.

[44] Fire is often God's instrument of judgement against the ungodly, Isaiah 29:6; 30:30; 66:15–16; Ezekiel 38:22; Amos 1:4, 7, 10, 12, 14; 2:2, etc.; Obadiah 1:18; Nahum 2:13.

Old Testament nowhere teaches that the whole cosmos is destined to be dissolved in fire.[45] In post-biblical Jewish eschatology, the notion is very rare; indeed, as Van der Horst points out, the only certain instances of it are some passages in the *Sibylline Oracles* (2:196–213; 3:80–92; 4:171–192; 5:155–161) in which Stoic influence is at work.[46] In Graeco-Roman cosmological discussion, the notion of a total cosmic conflagration was a characteristically Stoic idea. The author's depiction of the final judgement as a cosmic inferno is an intermingling of biblical prophecy and Stoic cosmology (prompted perhaps by Malachi 3:2–3 and 4:1, in which the day of the Lord is likened to fire).

In 3:6, the author portrays the Noahic flood as a total cosmic catastrophe which anticipates and parallels future conflagration. In doing so he seems to link the biblical/Jewish typological association of flood and final judgement with the Stoic notion of twin cosmic destructions by water and fire. The idea that the cosmos is subject to alternating destructions by flood and flame is not found in the early Stoics but is a feature of Roman Stoicism.[47] It is a projection onto a cosmic scale of the widespread belief in recurring local (though sometimes global) disasters by flood and fire.[48] It is also an attempt to correlate the beginning and end of world-orders more closely: the fire and water which dominated at creation gain the upper hand, in alternating catastrophes, at the end.

In 2 Pet 3:5–7, the author highlights three great moments in the history of the world, creation, flood, and God's final intervention in judgement, and matches these with key events in a Stoic view of cosmic history—the watery emergence of the world, the cosmic cataclysm and the great conflagration. He does so while avoiding the cyclic aspect of the Stoic scheme.

---

[45] *Contra* C.P. Thiede, 'A Pagan Reader of 2 Peter: Cosmic Conflagration in 2 Peter 3 and the *Octavius* of Minucius Felix', *Journal for the Study of the New Testament* 26 (1986), 79–96, esp. 80. Cf. Bauckham, *Jude, 2 Peter*, 300. Outside of 2 Peter 3, there are no explicit references to such an idea in the New Testament.

[46] P.W. van der Horst, '"The Elements will be Dissolved with Fire": The Idea of Cosmic Conflagration in Hellenism, Ancient Judaism and Christianity', in: P.W. van der Horst, *Hellenism, Judaism, Christianity: Essays on Their Interaction*, Kampen 1994, 227–251, esp. 243.

[47] Seneca, *Natural Questions* 3.27, 29 (in the latter, he ascribes it also to Berosos); *On Consolation (ad Marciam)* 26.6. Cf. Origen, *Contra Celsum* 4.64.

[48] E.g. Plato, *Timaeus* 22C–E. The idea of parallel (local and global) disasters by flood and fire is taken up by Jewish writers, e.g., Josephus, *Jewish Antiquities* 1:70; *Life of Adam and Eve* 49.3. Philo, *De vita Mosis* 2.53 applies the scheme to the Noahic Flood and the destruction of Sodom and Gomorrah. The writer of 2 Peter appears to do the same in 2:5–7.

5. *The argument from ἐξ ὕδατος καὶ δι᾽ ὕδατος in 2 Pet 3:5–7*

As noted at the beginning of the essay, the author's reference to the world's creation ἐξ ὕδατος καὶ δι᾽ ὕδατος forms part of a three-step argument for cosmic destructibility, involving ὕδατι in 3:6 and πυρί in 3:7. His line of reasoning may now be elucidated. Underlying the logic is the Stoic physical principle articulated by Seneca: 'Water and fire dominate earthly things. From them is the origin, from them the death.'[49] Verse 5 contains the first stage of the argument. Heaven and earth were formed out of the element water (ἐξ ὕδατος). Water, however, was not the earliest form of this element but the altered condition (δι᾽ ὕδατος) of a more pristine entity, pure fire. Since the ordered world arose from fire and water, it is destined to be resolved into these twin elements.[50] Verse 6 expresses the second point. The cosmos was destroyed by water (ὕδατι) at the time of the flood, when it returned to its primordial aquatic state. Verse 7 draws the conclusion derived from points one and two. The world now awaits a second destruction, this time by fire (πυρί). This destruction will take place at the parousia, when God will intervene decisively to judge the ungodly. The author thus argues for coming cosmic destruction from the physical origins of the world, informed by Stoic cosmogony. The world's emergence from fire and water points to twin cosmic catastrophes by these phenomena. The watery destruction lies in the past; a fiery destiny lies ahead.

6. *The appeal to Stoic cosmogony in the context of the author's debate*

The author's appeal to Stoic physical theory is an understandable apologetic move within the context of his debate with his opponents. The issue at stake in these verses is whether the cosmos is liable to destruction—a sub-issue in a debate about the likelihood of the parousia. The adversaries argued that the heavens and earth are not subject to a world-ending catastrophe; the universe is a secure and stable structure which will endure forever without real physical change (3:4). The writer seeks to show that the world is both capable of and destined for dissolution. This aspect of the controversy in 2 Peter 3 reflects a long-

---

[49] Seneca, *Natural Questions* 1.28.

[50] On the twin assumptions that generation implies destructibility and that all things dissolve into that out of which they were generated.

running philosophical dispute between Platonists and Aristotelians, on the one hand, and Stoics and Epicureans, on the other.[51] The former upheld the imperishability of the cosmos; the latter denied this belief. The author's opponents, it is clear, assume the position taken by Platonists and Aristotelians. The writer responds, therefore, by invoking the rival Stoic point of view.[52] This enables him to offer a scientific defence of his eschatological perspective as well as a biblical one.[53]

One of the main arguments advanced by Platonists and Aristotelians against the idea that the world will eventually pass away was a physical argument. It was given its classic form by Aristotle himself.[54] In order to be destroyed, he claimed, the universe must be subject either to external or to internal causes of destruction (or both). The first is inapplicable; nothing exists beyond the universe, so it cannot be affected by anything outside. The second cause is also excluded. If the world could be destroyed by something inside it, a part would be able to undo the whole, and this is contrary to all reason. Since no physical force can exert a lethal influence on the cosmos, he concluded, it will endure unceasingly. The Stoics replied by insisting that there *is* an internal force which can bring about total cosmic destruction—the element fire.[55] To the protestation that no part is strong enough to demolish the whole, they countered that during *ekpurosis*, fire actually *becomes* the whole. At the moment of conflagration, when the entire cosmos is ablaze, fire—the originating element—is once again *all* and *in all*. In later Stoicism water was also regarded as a part that periodically becomes the whole.[56]

---

[51] The best ancient account of this discussion is perhaps Philo's tract, *De aeternitate mundi*; Philo gives attention to the debate between Aristotelians and Stoics. On the question of the Philonic authorship of this treatise see D.T. Runia, 'Philo's *De Aeternitate Mundi:* The Problem of its Interpretation', *VC* 35 (1981) 105–151.

[52] A similar cosmological debate is held between the Stoicizing author of Ephesians and the Platonizing author of Colossians, according to George H. van Kooten, *Cosmic Christology in Paul and the Pauline School: Colossians and Ephesians in the Context of Graeco-Roman Cosmology*, Tübingen 2003, esp. chaps. 4.3.2 (b), 4.3.3, 4.5.3, 4.6.3, and 4.8.3.

[53] One may compare the appeal the second-century Christian apologist Minucius Felix makes, in *Octavius* 11.1–3, to Stoic physical theory in his defence of belief in the future destruction of the world by fire, in response to an opponent who rejects it as contrary to the theorem of cosmic indissolubility.

[54] The argument was presented in Aristotle's *On Philosophy*, now lost. Fragments of this work, however, are preserved in Philo's *De aeternitate mundi*. The physical argument appears in *De aeternitate mundi* 20–24.

[55] Mansfeld, 'Providence', 144–145.

[56] Seneca, *Natural Questions* 3.28.

There is no indication that the author's opponents raised Aristotle's physical objection to the notion of the destruction of the world; their rejection of it, as we have seen, was on empirical grounds. Yet, in 3:5–6, the writer does seem to addressing the issue: Whence comes the material for world-destruction? Perhaps he was aware that this issue was a traditional stumbling-block to acceptance of the notion of cosmic destructibility. The answer he implicitly gives to this question, as Calvin observed, is that the cosmos 'contains the material for its own ruin whenever it may so please God.'[57] By drawing on Stoic physical theory, he offers a neat rebuttal of Aristotle's contention that a part can have no power over the totality (whether our author intended to do so or not): when the world is destroyed by water or fire, it is not undone by a part, but reconstituted *wholly* such as it was before.

7. *Conclusion*

To sum up. The author's curious statement in 2 Pet 3:5 that heaven and earth were formed out of and through water can be explained as an attempt to correlate the Genesis notion of creation's watery origins with Stoic teaching about the world's beginnings. Adopting Stoic scientific theory, he maintains that the ordered cosmos derives from a single substance—the element water. Water was not the initial form of this substance, however, but a material transformation of a more primary pre-cosmic existent. He does not explicitly say what that existent was, but most probably, following Stoic cosmogony, he took it to be fire.

The author's appeal to Stoic cosmogony contributes to an argument for cosmic destructibility from the world's physical origins. The physical cosmos emerged from the element water which had previously been fire; it is thus fated to be resolved into water and fire, in parallel destructions. The world returned to water at the time of the flood; at present, it is reserved for fire. The author's utilisation of Stoic physics is appreciable in the context of a debate with opponents which mirrors the long-established philosophical debate on whether the cosmos endures without end or passes away.

[57] J. Calvin, *The Epistle of Paul the Apostle to the Hebrews and the First and Second Epistles of St Peter* (transl. W.B. Johnston; Calvin's Commentaries), Edinburgh 1963, 362.

THE HISTORY-OF-RELIGIONS BACKGROUND OF 1 TIMOTHY 4:4: 'EVERYTHING THAT GOD HAS CREATED IS GOOD'

Boudewijn Dehandschutter

Introduction

The student of early Christianity is, generally, familiar with the story of the Martyrs of Lyon and Vienne. This story is known to us through the *Ecclesiastical History* of Eusebius of Caesarea who quotes it at length in the first and second chapters of the fifth book.[1] What tends, again generally, to be less well known is that the 'Father of Church History' returns to the story of the Lyonese martyrs, even immediately afterwards in the third chapter of the fifth book! The text there runs as follows:

> Now the same document of the aforesaid martyrs contains also another story which deserves to be remembered, nor can there be any objection to my bringing it before the knowledge of my readers. It runs thus. There was one of them, a certain Alcibiades, who lived in absolute squalor, partaking formerly of nothing whatever save bread and water only; and he essayed to continue this mode of existence in prison also. But it was revealed to Attalus, after his first conflict in the amphitheatre was completed, that Alcibiades was not doing well in refusing the creatures of God and leaving an example whereat others might stumble. So Alcibiades was persuaded, and began to receive all things freely and give thanks to God. For they were not unvisited by the grace of God, but had the Holy Spirit for their Counsellor.[2]

There is no doubt that in the reaction of Attalus to the attitude of Alcibiades,[3] and in the conclusion of the passage, a reference can be

[1] Edition by E. Schwartz, *Eusebius Werke: Die Kirchengeschichte*, vol. 1, Berlin 1999 (2nd edn by F. Winkelmann).

[2] Translation by H.J. Lawlor and J.E.L. Oulton, *Eusebius: The Ecclesiastical History and the Martyrs of Palestine*, vol. 1, London 1927, 149.

[3] Alcibiades is mentioned only here in the quotations of Eusebius; Attalus appears more than once in the narrative, so that it is difficult to locate the passage from book V.3.

read to the text under discussion here: 1 Tim 4:4. The passage quoted might even be regarded as the first instance of reception of the verse and its context.[4]

One can discover a similar case slightly later in the *Adversus Haereses* of Irenaeus. When discussing several heresies at the end of the first book and pointing to that of the Encratites, the heresiologist states:

> the so-called Encratites, who sprang from Saturninus and Marcion, preached abstinence from marriage and so made void God's pristine creation, and indirectly reproved him who made male and female for generating the human race. They also introduced abstinence from what is called by them animal food, being thus *ungrateful* to the God who made all things ...[5]

As is well known, Irenaeus was eagerly quoted by Eusebius at the end of the fourth book (chap. 29), when introducing the heresies at the time of Marcus Aurelius.[6] One could easily continue the list of references to the verse in question in contexts in which an exaggerated asceticism leading to the rejection of marriage and food is rebuked. One other example, also well known, is Clement of Alexandria, *Stromateis*, book III. Clement does not fail to recall the position of the Apostle when dealing with the criticism of marriage by many ascetics.[7] Neither Ire-

[4] Cf. J. Allenbach, *Biblia Patristica: Index des citations et allusions bibliques dans la littérature patristique*, vol. 1: *Des origines à Clément d'Alexandrie et Tertullien*, Paris 1975.

[5] Irenaeus, *Adversus Haereses* I.26.1; translation by D.J. Unger *et al.*, *Saint Irenaeus of Lyons: Against the Heresies* (ACW 55), New York 1992, 93. 'Ungrateful' translates the Greek ἀχαριστοῦντες (cf. Eusebius, *Historia Ecclesiastica* IV.29.2), which, in our opinion, takes up the term εὐχαριστία of 1 Tim 4:4.

[6] On the treatment of heresies by Eusebius in his *Historia Ecclesiastica*, see R.M. Grant, *Eusebius as Church Historian*, Oxford 1980, 84–96.

[7] We refer in the following to the edition of O. Stählin, *Clemens Alexandrinus: Stromata Buch I–VI*, Berlin 1985^4; for an English translation, mainly of book III: J. Ferguson, *Clement of Alexandria: Stromateis, Books One to Three* (FCh 85), Washington 1991. There will be more than one occasion to refer to book III below, but some important issues can be highlighted in advance. Clement criticises Marcion and his followers continuously for their contempt of God's generous goodness (III.12–13 and 18.3–19.5), implying that Marcion misunderstood Plato (cf. 19.5 the quotation from Plato's *Statesman*: 'All that is good is got from the supreme disposer'); 22.1 criticising that Marcion's followers understood the Greek reluctance towards birth 'in a godless sense and show no gratitude to the Creator' (see further 34.3 and 38.2). Again the long passage 51–53 in which 1 Tim 4:1–3 is quoted; 63–73 with reference to heretics who 'attack God's creation under the pious name of self-control'; 80.3–81.1 against Tatian; 85.1–2 with again a quotation from 1 Tim 4:1–5, and the following consequence: 'it follows of necessity that there is no ban on marriage, or eating meat, or drinking wine'; 87 qualifying the claims about not marrying and its advantage with regard to the resurrection; 91 and the following sections against Julius Cassian and Marcion, to conclude: 'Under the name of what

naeus nor Clement allow any criticism of the good creation, because, 'everything that God has created is good, and nothing is to be rejected when it is taken with thanksgiving.'

But does all this shed any light on the reason why the post-Pauline author who wrote 1 Tim uttered his statement with such emphasis? Even if second-century authors receive the verse in a context of reaction to, and rejection of, exaggerated asceticism that implies a criticism of the goodness of creation,[8] this does not necessarily mean that *that* context is present for the author who wrote 1 Timothy. It is the purpose of the present article to elucidate why 1 Tim 4:4 has to take up the defence of 'everything that God has created.' The following discussion will investigate (1) the general context of the verse, i.e. the Pastoral Epistles to which it belongs;[9] it will try to decipher (2) the teachings so much detested by the author of 1 Tim, and try to identify more precisely (3) the background of 1 Tim 4:4.

1. *The Pastoral Epistles*

The Pastoral Epistles were written, one can argue, to defend and propagate Christianity as the true 'eusebeia' (εὐσέβεια). The Pastoral Epistles draw on the terminology and a number of ideas which belong to the Hellenistic religious discourse. To begin with, the term 'eusebeia' itself occurs with striking frequency in the Pastoral Epistles in comparison with other early Christian writings.[10]

1 Tim 3:16, moreover, reveals the truly great mystery of *eusebeia*: 'He who was manifested in the body, vindicated in the spirit, seen by angels; who was proclaimed among the nations, believed in throughout the world, glorified in high heaven.'[11] Other Hellenistic influences in the Pastoral Epistles include the emphasis on the notions of ἐπιφάνεια,

---

they falsely call knowledge, they have embarked on the road to outer darkness' (109.2). Elsewhere, too, Clement never misses an opportunity to defend God's creation.

[8] See Irenaeus and Clement, note 7 above.

[9] For introductory questions about the Pastoral Epistles, we refer to the three-volume commentary by L. Oberlinner, *Die Pastoralbriefe* (HTKNT XI.2), Freiburg 1994–1996.

[10] See εὐσέβεια in 1 Tim 2:2; 3:16; 4:7–8; 6:3, 5, 6, 11; 2 Tim 3:5; Titus 1:1; εὐσεβῶς in 2 Tim 2:12; Titus 2:12; and εὐσεβεῖν in 1 Tim 5:14. Cf. εὐσέβεια in Acts 3:12 and 2 Pet 1:3, 6, 7; 3:11; εὐσεβεῖν in Acts 17:23; and εὐσεβής in Acts 10:2, 7; 2 Pet 2:9. And see W. Foerster, 'Eusebeia in den Pastoralbriefen', *New Testament Studies* 5 (1959) 213–218.

[11] Translation from the *New English Bible*.

φιλανθρωπία, the ideas about God as σωτήρ, saviour, and at the same time very spiritualised descriptions of God as ἄφθαρτος and ἀόρατος (1 Tim 1:17), as alone having immortality, ἀθανασία, dwelling in an unapproachable light, the One whom no one has seen or can see (1 Tim 6:16). To these one can add the connection between religion and ethics, in the context of *eusebeia* as religion with its public function. Thus the Pastoral Epistles propagate respect for civil authorities: emperors and rulers deserve Christian prayers, so that we can lead a quiet life, in all decent piety (1 Tim 2:1–2; cf. Polycarp, *To the Philippians* 10/12). This behaviour is good and agreeable to our Saviour God who wants all men to be saved and to come to the cognition of truth (2:3–4).[12] In this light, it is perhaps no exaggeration to consider the Pastoral Epistles as an attempt to make the Christian religion respectable and acceptable to a non-Christian, pagan environment (cf. 1 Tim 3:7).[13]

This would also explain the author's concern about women and slaves behaving within the limits of social acceptability: a woman should remain in quiet subordination, not have the notion to teach, but should excel in what πρέπει, what is fashionable (1 Tim 2:10).[14] The whole 'Haustafel'-instruction aims at the same goal: Christians are εὐσεβεῖς, not endangering the socio-religious order. Of course this is not possible without good leadership in the communities. The moral and intellectual capacities of Christian leaders have the full attention of the author, but the true context for this concern should be apparent from the above.[15] The concern about good leaders goes together with a concern about proper teaching (as these leaders should, above all, teach the right διδασκαλία). Timothy himself is to take care of the reading, the admonition and the teaching[16] 'until Paul arrives' (1 Tim 4:13). Certainly, proper teaching is not a matter for neglect, because doctrinal

[12] This certainly contrasts with the 'knowledge falsely so-called' in 1 Tim 6:20.

[13] See also the passages on 'Christian in the world' and the excursus on 'godliness' in the excellent new commentary by R.F. Collins, *1 & 2 Timothy and Titus: A commentary*, Louisville & London 2002, 54–59; 122–126.

[14] Cf. M.R. D'Angelo, 'Eusebeia: Roman Imperial Family Values and the Sexual Politics of 4 Maccabees and the Pastorals', *Biblical Interpretation* 11 (2003) 139–165.

[15] Contemporary exegesis also had a tendency to focus almost exclusively on the notion of Christian leadership, not to mention ecclesiology, as *the* topic of the Pastorals. Cf. the long excursus in L. Oberlinner, *Die Pastoralbriefe: Titusbrief*, Freiburg 1996, 74–101, 74: 'das allesbeherrschende grosse Thema'.

[16] The notion of teaching has an overall presence in the Pastorals; it is important to recognise that this intellectual aspect of leadership is pre-eminent. It supports my view of the Pastorals as a kind of 'propaganda' literature.

ambiguities might make the Christian *eusebeia* suspicious or detestable. It is not at all surprising, then, to encounter, in the very first verse of 1 Tim after the address, the notion of ἑτεροδιδασκαλεῖν (cf. 1 Tim 1:3, and again 6:3). Paul entrusts Timothy with the task of staying on at Ephesus and putting an end to such activities! Let us now turn to this ἑτεροδιδασκαλεῖν as the concrete context for 1 Tim 4:4.

2. *The 'false' doctrine in the Pastoral Epistles*

In order to gain an image of the teachings viewed as 'insane' by the author of 1 Tim, we should not concentrate only on the latter epistle. If one approaches the Pastoral Epistles as a pseudepigraphical *corpus*, one must also take 2 Tim and Titus into consideration.[17] For the author of the Pastoral Epistles, 'doctrinal' deviation is no light matter: at the end of Titus we find the warning about a heretical person. It is not merely a question of divergences of opinion: the αἱρετικὸς ἄνθρωπος should fundamentally change his mind, or be thrown out (Titus 3:10). However, the reconstruction of the ideas considered so dangerous is not easy. One can collect the following elements:

a. The 'adversaries' are charged with inclining to myths and genealogies, even from the very beginning (1 Tim 1:4). These myths are characterised elsewhere as 'Jewish' (Titus 1:14), and one is inclined to compare this with Ignatius, *To the Magnesians* 8.[18]
b. In 1 Tim 6:20, at the very end of the letter, it is said that Timothy should adhere to that which has been committed to him (παραθήκη) and reject the stupid idle talk and the contradictions of what is falsely called 'knowledge' (γνῶσις). This sentence has made its way into heresiological literature; it is difficult to avoid the impression that a doctrinal 'system' is being denounced here.[19]

---

[17] Cf. the commentaries by Oberlinner and Collins, as well as R. Burnet, 'La pseudépigraphie comme procédé littéraire autonome: l'exemple des Pastorales', *Apocrypha* 11 (2002) 77–91.

[18] Cf. Ignatius, *To the Trallians* 5; for the interpretation of the passage *To the Magnesians* 8, see e.g. P.Th. Camelot, *Ignace d'Antioche: Polycarpe de Smyrne; Lettres; Martyre de Polycarpe* (SC 10), Paris 1969[4], 86 note 3: 'Docétisme judaïsant ... teinté de gnose'.

[19] Cf. G. Haufe, 'Gnostische Irrlehre und ihre Abwehr in den Pastoralbriefen', in: K.W. Tröger (ed.), *Gnosis und Neues Testament: Studien aus Religionswissenschaft und Theologie*, Gütersloh 1973, 325–339.

c. A direct and well known statement is that referred to in 2 Tim 2:18 as the aberration of Hymenaios and Philetus that the resurrection has already taken place. If the Pastoral Epistles have some connection with an early form of gnosis (as many commentators have assumed and continue to assume), this statement could be considered a most explicit element of the 'heresy', having been documented now by more than one text from the Nag Hammadi Library.[20]
d. A fourth indication is provided by the text of 2 Tim 3:6–7, which speaks of people who intrude into houses and try to convince women, who always want to learn and never arrive at the recognition of the truth. This is thought to point to the success of gnostic teachers with regard to women, and simultaneously to explain the call for subordination in texts such as 1 Tim 2:9–15 (cf. 5:14).[21]
e. A final indication of the opponents' ideas is given by the central passage under discussion: they reject marriage and particular foods (1 Tim 4:4). This problem is also alluded to in Titus 1:14–15 (at least according to some explanations) as a failure to recognise the goodness of creation. Some exegetes, and I am inclined to follow them, want to connect this with the sudden admonition given by 'Paul' to Timothy not to drink only water but also some wine (1 Tim 5:23). In this way Timothy could avoid making the impression of adherence to a strict asceticism (that also rejected the use of wine). We know from Clement of Alexandria (*Paedagogus* II) that there was a need to defend wine as an element of God's good creation, which was in no way objectionable in itself if taken in moderation.[22] As a model of a Christian leader, Timothy must

[20] It is beyond the scope of the present article to consider more appropriate texts such as the *Epistle to Rheginus*, the *Gospel of Philip*, the *Exegesis on the Soul*, or the *Testimony of Truth*, all of which are Nag Hammadi tracts which discuss the notion of resurrection. Although F. Wisse, 'The "Opponents" in the New Testament in the Light of the Nag Hammadi Writings', in: B. Barc (ed.), *Colloque International sur les textes de Nag Hammadi (Québec 22–25 août 1978)*, Québec & Louvain 1981, 99–120, esp. 108–114, has doubts about the identification of the opponents in the Pastoral Epistles with adherents of the ideas of the Nag Hammadi texts, I would maintain that there might be a connection with the ideas present in e.g. the *Testimony of Truth*. See also A. and P. Mahé, *Le Témoignage veritable (NH IX,3): Gnose et martyre*, Québec & Louvain 1996.

[21] For the enormous influence of these texts on later theology, see e.g. R. Nürnberg, '"Non decet neque necessarium est, ut mulieres doceant": Überlegungen zum altkirchlichen Lehrverbot für Frauen', *Jahrbuch für Antike und Christentum* 31(1988) 57–73; and E.M. Synek, 'In der Kirche mögen sie schweigen', *Oriens Christianus* 77 (1993) 151–164.

[22] On the anti-Encratite inspiration of this text of the *Paidagogos*, see B. Dehand-

beware of the dangers of intemperate drinking of wine, see 1 Tim 3:3 etc., but he should not be identified either with people whose abstinence is motivated by a criticism of creation.[23]

3. *The history-of-religions background*

Closer identification of the 'adversaries' against whom 'Paul' is warning[24] is not easy, and depends on the way in which one groups the elements of the doctrine scattered through the Pastoral Epistles. There is a common tendency to focus on only one of them. Another danger in the identification of the 'false' teachers, is to situate them immediately in the line of *heretics*—at least people who are considered as such—who are more or less contemporary, such as those presumably behind the warnings of the Ignatian Epistles and/or Polycarp's *Letter to the Philippians*. This is the approach, for instance, of K. Rudolph in his classic book on gnosis, according to which the Ignatian Epistles also refer to some gnostic 'failures', docetism and the denial of the resurrection, and as a consequence argue in favour of a strong leadership that guarantees unity in the church.[25] However this may be, I am inclined to believe that the situation of the Pastoral Epistles is still different, at least in the sense that the doctrinal questions can not be identified: the Pastoral Epistles make no allusion to the problem of docetism, and the statement about the *denial* of the resurrection (in Polycarp, *To the Philippians* 7) should not be identified with the claim that the resurrection has already taken place, as referred to in 2 Tim 2:18.

We can do no better than to take the verse under discussion here and its context as a guideline. What does it say? Having eulogized the mystery of the *eusebeia* (1 Tim 3:9, 16), the author warns solemnly

schutter, 'Mèketi hudropotei: Some notes on the Patristic Exegesis of 1 Timothy 5:23', *Louvain Studies* 20 (1995) 265–270.

[23] Cf. the texts below from Clement, *Stromateis* III.45, 48, as well as M. Wolter, *Die Pastoralbriefe als Paulustradition*, Göttingen 1988, 256–270, on the question of the adversaries. See also G.C. Streete, 'Askesis and Resistance in the Pastoral Letters', in: L.E. Vaage and V.L. Wimbusch (eds), *Asceticism in the New Testament*, London & New York 1999, 299–316.

[24] As a rule the difference between 'adversaries' and 'heretics' is overlooked in this case.

[25] K. Rudolph, *Die Gnosis: Wesen und Geschichte einer spätantiken Religion*, Göttingen 1978, 321–322.

against people, inspired by erroneous spirits and demonic teachings,[26] who forbid marriage and promote abstention from particular food (4:1–3b). In the author's view, these things are created by God and should be enjoyed with thanksgiving (4:3c–4). Since 1 Tim 4:4 states that God's creation is good and nothing should be rejected, the discussion about *creation* is shown to be an important topic, much more than is usually recognised. This discussion is perfectly in line with the reference to myths and genealogies, which can be taken as an allusion to gnostic stories on creation insisting on the failures of the demiurgical powers. 1 Tim 4, like several other (later) Christian authors (cf. Irenaeus, Clement), recognizes that exaggerated asceticism often implies a criticism of creation. With his insistence on the goodness of creation, the author of 1 Tim seems to echo the repetitive statements in the creation account of Gen 1, that God saw that it was good (Gen 1:10, 12, 18, 21; cf. 1:31 and 1:4; see also Noort, this volume, § 1).

To put it for a moment in Clement's words: 'What about those who use religious language for irreligious practices involving abstinence against creation and the holy Creator, the one and only almighty God, and teach that we ought not to accept marriage and childbearing …' (*Stromateis* III.45). Refuting these people (but one could just as well quote the explicit reaction against Marcion and his followers earlier in the third book, chap. 12), Clement makes an interesting observation (chap. 48): 'If as they claim, they have already attained the state of resurrection, and for that reason repudiate marriage, they should *stop* eating and drinking …' It is as if the Alexandrian is ridiculing the teachers referred to in 1 Tim 4:4: their asceticism is not consistent enough for spiritual people.[27] Clement reminds us that the claim about the resurrection is not foreign to this context: radical asceticism goes together with the consciousness of a spiritual, pneumatic status.[28]

To return to the Pastoral Epistles, if we consider them as a whole, it does not seem appropriate to approach the question of exaggerated asceticism in 1 Tim 4 on the level of a criticism of ascetical behaviour inspired by rules of *purity*, as seems to be the case in Rom 14, or on the level of alimentary temperance, which is what seems to be at issue in Ephesians 5:18, a text warning about the abuse of wine, which in

[26] 1 Timothy anticipates the demonological deduction of heresy in later authors, such as Tertullian, *De Praescriptione Haereticorum* VII and Eusebius, *Historia Ecclesiastica* IV.7.

[27] Cf. Clement, *Stromateis* III.60.1 and III.87.

[28] But see the answer of Clement, *Stromateis* III.51.3.

the history of reception has often been combined with 1 Tim 5:23.[29] No doubt, early Christianity did invite ascetic behaviour. However one conceives the problems that are present in Colossians 2, the practice which underlies Col 2:16ff. offers an example of how some Christians regarded radical ἄσκησις as the appropriate way of life. Apart from a number of cases in the Pauline tradition, one also cannot escape the conclusion that in the early transmission of the Sayings of Jesus, even in the earliest stage of that transmission (i.e. the sayings on following and perfection), the message of a radical asceticism is present.[30] Clement of Alexandria makes clear that this interpretation of the sayings of Jesus has taken on dimensions that need an urgent defence of marriage and procreation (*Stromateis* III).[31]

What is striking in the context of 1 Tim 4:4, however, is that the whole question is explicitly connected with *creation*—this is to be accepted as good, i.e. the work of a good Creator. So we are invited to investigate who questioned all this, at the time of the writing of the Pastoral Epistles. There is a growing tendency to speak about early gnosticism as the background of the text of 1 Tim 4.[32] I am inclined to follow this, but not without raising the question: what is early gnosticism? And what does it mean in relation to the time of writing of the Pastoral Epistles?

Let us take the latter question first. The discussion on this point must be brief, but there is nevertheless reason to make a double statement. (1) It is sufficiently certain that Polycarp's *Epistle to the Philippians* is a testimony, the earliest one, to the reception of the Pastoral Epistles. This has been convincingly demonstrated by two recent monographs, those by Berding and Hartog.[33] (2) The date of Polycarp's *To the Philippians*

---

[29] Cf. Dehandschutter, 'Mèketi hudropotei', as in note 22 above.

[30] This is not only true of the materials contained in the *Gospel of Thomas* but also of other Nag Hammadi texts. See H.M. Schenke, 'Radikale sexuelle Enthaltsamkeit als hellenistisch-jüdisches Volkommenheitsideal im Thomasbuch (NH II,7)', in: U. Bianchi (ed.), *La Tradizione dell'Enkrateia: Motivazioni ontologiche e protologiche*, Rome 1985, 263–291; R.McL. Wilson, 'Alimentary and Sexual Encratism in the Nag Hammadi Tractates', in: Bianchi, *La Tradizione dell'Enkrateia*, 317–339.

[31] In the same sense, Clement does not reject the *Gospel of the Egyptians*, but the abuse made by the Encratites of the sayings of the Lord it contains.

[32] See the commentary by Oberlinner, *Die Pastoralbriefe*.

[33] The position of H. von Campenhausen, 'Polycarp von Smyrna und die Pastoralbriefe', in: Idem, *Aus der Frühzeit des Christentums: Studien zur Kirchengeschichte des ersten und zweiten Jahrhunderts*, Tübingen 1963, 167–252, that Polycarp himself might be the author of the Pastoral Epistles is no longer accepted. See K. Berding, *Polycarp and Paul: An Analysis of their Literary and Theological Relationships in Light of Polycarp's Use of Biblical and*

can be put at 120 or even 115;[34] this means a *terminus ante quem* for the Pastoral Epistles, leading to a date of writing around the turn of the century, let us say 100 CE. What about early gnosticism? Whatever we think about this phenomenon in early Christianity, we can agree that even for its earliest representatives *creation* was a crucial matter.[35] In their view, creation is *not* the activity of the most high God, the uncreated Father,[36] but is due to the intervention of powers, angels or archons, and by that fact detestable. Early gnostics were led to their motivation of ascetism by their depreciation of the creational activities of demiurges or archontes as responsible for this world.[37]

What now is the earliest information we can get about this? It is not impossible to find at least a hint for the context of 1 Tim 4:4 in Irenaeus, *Adversus Haereses* I.24. Irenaeus presents the 'doctrine' of Satornil (Saturninus), adding that this heretic considers marriage and procreation as coming from Satan, and concludes that the followers of Satornil as a majority abstain from ἔμψυχα, 'animalibus', and mislead many by this form of feigned abstinence: 'Besides, he said that to marry and get children comes from Satan. Most of his followers even abstain from animal food, misleading many by this false type of temperance'.[38]

Can we connect this text directly with 1 Tim 4:4? All I wish to point out is that this verse can illustrate that, at the time of the writing of the Pastoral Epistles, there were (Christian) teachers who understood creation in a negative sense, with all the ascetic consequences that

---

Extra-Biblical Literature, Leiden 2002; and P. Hartog, *Polycarp and the New Testament: The Occasion, Rhetoric, Theme and Unity of the Epistle to the Philippians and its Allusions to New Testament Literature*, Tübingen 2002.

[34] The latter date is Hartog's; 120 CE is a date accepted by many contemporary scholars, including Berding. But Berding follows the common opinion on the division of the Polycarpian letter, whereas Hartog does not. See my own position in B. Dehandschutter, 'Polycarp's Epistle to the Philippians: An Early Example of "Reception"', in: J.M. Sevrin (ed.), *The New Testament in Early Christianity*, Louvain 1989, 275–291; and my reaction to Berding in: *Vigiliae Christianae* 58 (2004) 98–101

[35] It is difficult to understand why the theme of creation is lacking in W. Foerster's 'Hauptmomente der Gnosis'. Cf. W. Foerster, *Gnosis: A Selection of Gnostic Texts*, vol. 1: *Patristic Evidence*, Oxford 1972, 9.

[36] See B. Dehandschutter, 'Théologie negative: la contribution des textes gnostiques et hermétiques', in: M.M. Olivetti (ed.), *Théologie négative*, Milan 2002, 505–513.

[37] Cf., by way of contrast, libertism, which is probably an issue in the Epistle of Jude. See also the discussion by the present author in: J.J.A. Kahmann and B. Dehandschutter, *De tweede brief van Petrus: De brief van Judas*, Boxtel 1983, 113–150.

[38] Translation of D. Unger, *Saint Irenaeus of Lyons*, 84–85; see also the commentary, 230–231; and H. Schlier, 'Das Denken der frühchristlichen Gnosis (Irenäus Adv. Haer. I 23.24)', in: *Neutestamentliche Studien für Rudolph Bultmann*, Berlin 1954, 67–82.

entails. To put it differently, earlier than Basilides, Valentinus, Marcion and others, even earlier than Satornil, Christians had been thinking about creation. But despite what one would expect, they did not always do so in a positive sense, unlike the author of 1 Clement who describes the *harmony* of creation (in terms close to Stoic cosmology[39]). The author of the Pastorals came across people who held ideas about creation he could not agree with.

In addition to this interpretation, the remark might be made (in the context of 1 Tim 4:6–10, the verses succeeding the central passage under discussion)[40] that it is important to see that the author of the Pastoral Epistles links the words about the *good* creation immediately to his notion of *eusebeia*: the *Christian eusebeia* which is not concerned with profane, superstitious myths which might inspire stupid thoughts about creation and come into opposition with the sound teaching handed over to and by Timothy.

Elsewhere again, in 2 Tim 3:14–16, the insistence on Timothy's knowledge of Scripture, of divinely inspired Scripture, is again significant in counterbalancing myths and genealogies. And *eusebeia* not only implies the life to come, but also *this* life, so we should live *well* in this aeon (see 1 Tim 4:8 and Titus 2:12). To live εὐσεβῶς in this aeon means not to detest it. The many admonitions on Christian conduct throughout the Pastoral Epistles may all be read in this perspective. The author does not refrain in the following verses from advising young widows to marry, to raise children, to keep their households (5:14), contrary to what those ascetics would argue. It is the 'unjustified' asceticism of such people that gives rise to criticism and slander. And, as already pointed out, the 'apostle' himself gives the advice to Timothy to drink wine, not as a πάροινος, but as a σώφρων! Again, *eusebeia* implies the belief in a living God who is the saviour of all men. God is not the σωτήρ just of a privileged group of people (i.e., the pneumatics, or the two categories of men in the teaching of Satornil), but of all, and certainly of all believers.

The author of the Pastorals made an effort to defend the good creation; indeed Basilides, Valentinus and others were still to come, but the earlier author fired a warning shot in the direction of some gnosticising Christian teachers of his time: 'everything that God has created is good.'

---

[39] Cf. H. Lona, *Der erste Clemensbrief* (KAV 2), Göttingen 1998, 249–274.

[40] As a matter of fact the pericope 1 Tim 4:1–5 is too often separated from the following verses 6–10 in editions, translations and commentaries.

PART IV

CREATION IN THE MIDDLE AGES AND MODERNITY

READING CREATION: EARLY MEDIEVAL VIEWS OF GENESIS AND PLATO'S *TIMAEUS*

Willemien Otten

1. *The Book of Genesis:* opera aperta *or Christian classic*

Creation is generally considered a distinguishing feature of Christianity. When unpacking this concept, however, one soon notices how it displays an intrinsically loaded character. Ranging from medieval cosmological debates to modern discussions on intelligent design, creation is more than a foundational tenet of the Christian religion. Centuries of reading Genesis have produced endless subtexts, suggesting on the one hand ever new possibilities of 'reading creation', while on the other hand critically assessing their plausibility. The sheer diversity of these subtexts makes clear to us that creation somehow both anchors and confirms the uniqueness of the Christian world view. Creation accounts can be of an exegetical nature, for example, as shown by the various attempts to discriminate between the literary depiction of biblical creation and the epic strife of the gods in the Gilgamesh.[1] Or they may have a cosmological purpose, as when biblical creation is used to criticize contemporary science, with modern creationist accounts enforcing the former correspondence between the Book of Nature and the Book of Scripture.[2] In contemporary neo-orthodox theology, finally, a

[1] Cf. A. Heidel, *The Gilgamesh Epic and Old Testament Parallels: A Translation and Interpretation of the Gilgamesh Epic and Related Babylonian and Assyrian documents*, Chicago 1949[2].

[2] On the medieval use of this trope and its demise, see W. Otten, 'Nature and Scripture: Demise of a Medieval Analogy', *Harvard Theological Review* 88 (1995) 257–284. The twelfth century was in many ways the high point of this trope, but while its importance declined, the way the Bible was read philologically in tandem with nature clearly stimulated scientific scholarship, see P. Harrison, *The Bible, Protestantism and the Rise of Natural Science*, Cambridge 1998. On present day creation debates, see K. Doyle Smout (ed.), *The Creation/Evolution Controversy: A Battle for Cultural Power*, Westport, CT 1998; J.A. Moore, *From Genesis to Genetics: The Case of Evolution and Creationism*, Berkeley 2002; Robert T. Pennock (ed.), *Intelligent Design Creationism and Its Critics: Philosophical, Theological, and Scientific Perspectives*, Cambridge, MA 2002.

renewed polemical impulse puts the creation story in opposition to natural theology.[3] With a literary term from Umberto Eco, one is inclined to consider Genesis an *opera aperta*.[4]

The wide range of contemporary readings forms a radical contrast with the reading developed long ago by Augustine of Hippo. The latter's chief aim, borne out especially in his anti-Manichaean writings, was to develop an orthodox reading of Genesis corresponding with a sound philosophy of creation.[5] While he lacks the aggressive tone with which contemporary neo-orthodox theologians heighten the stakes for divine transcendence, he clearly closes off some of the avenues for prebiblical speculation found in Philo, Origen, and Ambrose. Choosing his own intellectual path, Augustine wanted to overcome intellectual strife by absorbing all cosmological disagreements into a new and overhauled Christian reading of Genesis.[6] It was to be literal rather than allegorical, conforming simultaneously to standards of scientific soundness and theological truth. For Augustine, then, Genesis was not so much an *opera aperta* but rather a Christian classic, as defined by David Tracy.[7] Although the soteriological focus characteristic of reformation exegesis is still absent, Augustine is much less speculatively inclined than his Platonic predecessors. As he aimed to find a balance between an appropriate anthropological focus and sound cosmology, the opening chapters

[3] See esp. Karl Barth, *Church Dogmatics* III.1: 'The Work of Creation'. For a brief discussion of Barth's covenantal and Christological reading of creation, see P. Fulljames, *God and Creation in Intercultural Perspective: Dialogue Between the Theologies of Barth, Dickson, Pobee, Nyamiti and Pannenberg*, Frankfurt am Mein 1993, 11–35.

[4] See U. Eco, *The Open Work* (transl. A. Cancogni), Cambridge 1989, 4: 'A work of art, therefore, is a complete and *closed* form in its uniqueness as a balanced organic whole, while at the same time constituting an *open* product on account of its susceptibility to countless different interpretations which do not impinge on its unadulterable specificity. Hence, every reception of a work of art is both an *interpretation* and a *performance* of it, because in every reception the work takes on a fresh perspective for itself'.

[5] See e.g. Augustine's *De natura boni*, dated around 404 and often seen as the last of his anti-Manichaean works, in which he connects his belief in a good and providential God with an orthodox exposition of *creatio ex nihilo*. Augustine's disenchantment with Manichaeism resulting from his rejection of their leader Faustus as a flawed scientist is described in *Confessions* V.3–7 (transl. R.S. Pine-Coffin), Harmondsworth 1961, 92–99.

[6] Augustine's interest in creation cannot be separated entirely from his development of a more pessimistic anthropology in his anti-Pelagian phase, as he had been driven before by a similar opposition to gnosticism. See E. Pagels, *Adam, Eve and the Serpent*, New York 1988, 98–150.

[7] David Tracy has defined classics as 'those texts, events, images, persons, rituals and symbols which are assumed to disclose permanent possibilities of meaning and truth'. See his *The Analogical Imagination: Christian Theology and the Culture of Pluralism*, New York 1981, 68, see further 107–115.

of Genesis made him aware of the need to erect certain boundaries by which to rule out unbridled speculation. His was not a natural theology, therefore, but an ultra-natural theology.

Yet creation also had sacramental meaning for Augustine, as early Christian theology sees an overlap of the doctrinal and the material, the symbolic and the liturgical. The story of creation does not just tell us about the world but informs us also about the community, i.e., the church, that was to live and enjoy that world. Just as the spirit of God graced the primal waters of creation, so for Tertullian it also inspired the waters of baptism.[8] If it is indeed true that Christianity's resounding cosmological grammar penetrated even the innermost recesses of its sacramental mysteries, this reflects Genesis' collective appropriation in the world of the Christian churches. Cosmos and community had become linked by a unique text which was itself taking on sacramental meaning. From Tertullian's *De baptismo* onwards, through the early Christian *Hexaemeron*-tradition and the cosmological exegesis of the early Middle Ages, the opening narrative of Genesis has featured prominently in the definition of Christian self-identity. With the clericalization of the church after the Gregorian reform and the concomitant professionalization of theological training in the twelfth century, however, the bond between liturgical celebration and sacramental definition grew increasingly tenuous.[9] In a movement that may reflect a similar reification of language, attempts to arrive at a universally accepted interpretation of creation also ended in failure. How the early medieval story of creation unfolded to peak in a disjointed and 'dual' reading of Genesis will be the subject of the following sections.

2. *Concept and process: the Platonic grammar of early medieval creation*

It is tempting to trace the fault-lines of this 'dual' resonance of creation, divided into a cosmological and a sacramental one, back to the historical differences between Platonism and Christianity. I shall briefly discuss the presuppositions of each system's approach to the origin of the

---

[8] In *De baptismo* IV Tertullian describes as the primary principle of baptism 'that the Spirit of God, who hovered over (the waters) from the beginning, would continue to linger over the waters of the baptized.' Cf. P. Cramer, *Baptism and Change in the Early Middle Ages c. 200–c. 1150*, Cambridge 1993, 52–63.

[9] See Cramer, *Baptism and Change*, 221–266 (The twelfth century, or falling short).

world to bring out how these two model interpretations are profoundly at odds with each other.

In explaining the origin of the world, the crucial problem for Platonism is how to overcome the tension between the One and the Many. Plato's works present us with various solutions. In the *Parmenides* (1) the transcendence of the One is such that it is ultimately declared irreconcilable with the concept of Being. With the One transcending the world, Being marks the transition to the lower world of pluriformity. With the noetic world rising far above the material world of humans, animals and plants, Proclean Neoplatonism turned the realm of transcendence into a complete spiritualized hierarchy. As an alternative solution, Plato's *Timaeus* (2) presents us with the image of a craftsman who creates a world in which plurality is key, matter and the elements the stuff from which it is made, while the soul acts as its principle of life. This solution most resembled the Christian view of creation in the eyes of twelfth-century philosophers and theologians.

With the Good, the True, and the Just seen as valid descriptions of divine transcendence, Plato's theory of Forms allows for a certain degree of equivocation. Reflecting the polysemy of the *Cratylus* (3), this theory of Forms was preserved in the sixth-century theory of Divine Names developed by Dionysius the Areopagite, which was adopted by Eriugena in his ninth-century creation epic *Periphyseon*.[10] While remaining important on the level of linguistic predication and mystical theology, however, the tension between apophatic and kataphatic predicates for God soon after ceased to influence medieval creation theories.

There seems to be an inherent contradiction between the grammar of Platonism, in the three forms listed above, and the Christian reading of Genesis. Is not the idea of *creatio ex nihilo* in sharp contrast with the Timaean formation of the world out of the four material elements: fire, water, earth, and air? Informing the creation of heaven and earth, and of all other aspects of the material world, the latter view appears premised on the acceptance of an essential correspondence between macro- and microcosm. If heaven and earth are made of the four elements, so human beings must be also, which undermines the Christian

[10] On the unique position of Eriugena as mixing Augustine's and Dionysius' notion of Forms, see St. Gersh, 'Cratylus mediaevalis—Ontology and Polysemy in Medieval Platonism (to ca. 1200)', in: J. Marenbon (ed.), *Poetry and Philosophy in the Middle Ages: A Festschrift for Peter Dronke*, Leiden 2001, 79–98, esp. 85–88.

idea of humanity as *imago Dei*. With the *Timaeus* serving as the main vehicle for medieval views of creation due to Calcidius' Latin translation and commentary, and the cosmos increasingly regarded as an animal, or a living organism with a soul, the analogy between the world and humanity began to lose much of its threat. Perhaps it could even have a desirable effect. For with humanity embedded in the whole of creation, the entire world—from the upper air to the netherworld—suddenly appears to bathe in divine light.[11] If we accept a Trinitarian view of the creator, moreover, allowing for a plurality of divine persons, the problem of the Platonic demiurge as a separate deity may also be overcome. Again the creation of humanity is the test case which must prove the claim of cosmic continuity. Using the plural to bring out the contribution of all of three divine persons, Genesis purposely states: 'Let us make man in our image and likeness …'[12]

Through these and other harmonizing tendencies, some of the problems that had plagued the interpretation of Genesis before, evaporated. In the ninth century, John the Scot Eriugena seemed still close to the patristic tradition, as his *Periphyseon* advocates a double creation of humanity. Dating back to Philo and mediated through Gregory of Nyssa, his interpretation of the Hebrew *parallellismus membrorum* of Gen 1:27 ('Let us make man in our image and likeness,' followed by 'man and woman he created them') displays an ontological hierarchy of the sexes. God first created an archetypal, if not androgynous human being, only to counteract on the disastrous effects of the fall by substituting Adam and Eve as physical human beings. While their sexually differentiated bodies testify to humanity's sinful state, they also allow for the new option of physical procreation.[13] Somehow or other, the

[11] Alan of Lille's famous stanza: 'Omnis mundi creatura / quasi liber et pictura / nobis est et speculum' is often taken as representative of this sacramental view of nature. For a more critical approach to this stanza in light of the entire poem, see Otten, 'Nature and Scripture'.

[12] So Peter Abelard, whose *Expositio in Hexaemeron* (PL 178), 760C–D posits that an internal discussion between the three divine persons preceded humanity's creation.

[13] In this way, procreation became a 'merciful afterthought' in Gregory's view. See P. Brown, *Body and Society: Men, Women, and Sexual Renunciation in Early Christianity*, 285–304, esp. 294. Eriugena basically adopted Gregory's position, although he no longer held that an actual spiritual creation preceded the creation of historical humanity based on the difference between *ante et post peccatum*. His famous comment 'homo melior est quam sexus' (*Periphyseon* II 534A) indicates that the adjectives male and female do not apply to human nature, but only to its division. This division will ultimately be undone by the undivided human nature of the risen Christ.

arrival at a perfect number of souls to secure the option of human salvation no longer puzzles twelfth-century commentators.[14] Replacing it as the age's central concern is the larger underlying question how best to express the continuity of macro- and microcosm through an integral concept of creation.

3. *Allegorizing creation*

To explain the increasing complexity of earlymedieval creation, we sometimes find the following view. Throughout the early Middle Ages, Genesis apparently functioned as the norm for describing the world's origin. When in the twelfth century the *Timaeus* is studied anew with fresh energy, it elicits a mostly scientific response, thereby tempting Christian thinkers to embrace unorthodox viewpoints. The latent tendency towards intellectual subversion is strengthened by the use of hermetic sources like the *Asclepius*, as a result of which the concept of biblical creation loses ground fast. Shedding excessive fideism, allegory makes the interpretation of creation palatable for the scientifically minded, preventing them at the same time from sliding into heterodoxy.

According to this schematized perspective,[15] the allegorical escape can only last as long as natural science is an underdeveloped discipline. With intellectuals beginning to take a natural science approach to creation, the allegorical reading of Genesis loses much of its attraction. The demise of allegory and the surrender of exegetical cosmology to natural science thus go hand in hand. With science involving recourse to reason, the literal reading of Genesis cannot fail to gain priority, even though in a sacramental countermove, Hugh of St. Victor and others engage in literal exegesis precisely to keep a strong focus on human salvation by linking creation and restoration.[16]

---

[14] It famously plagued Christian thinkers from Augustine's *City of God* to Anselm of Canterbury's *Why God Became Man*, leading to all sorts of speculations about human beings taking the place of fallen angels in the City of God, and especially about the question how many human beings could be accommodated. Remarkably, Eriugena seemed not bothered.

[15] Centring on the idea that twelfth-century scholars were primarily interested in Christianizing Plato, H. Fichtenau dwells on the intrinsic contradiction of Platonism and Christianity in his *Heretics and Scholars in the Middle Ages, 1000–1200* (transl. D. Kaiser), University Park, USA 1998, 172–196 (The Philosophical Myth: Platonists).

[16] See below note 48.

Although there is truth to the above scenario, and interest in literal exegesis certainly spreads in the twelfth century, it is by no means clear that the debate on orthodoxy and heresy has its roots either in the difference between Christianity and Platonism, or in that between faith and science.[17] Such schematic reconstructions of twelfth-century development seem too influenced by the modernist debate, forcing theology and exegesis to retire increasingly to the faith-camp. To come to a more nuanced perspective, three further arguments relating to the theme of 'reading creation' must be brought to bear on the discussion, circling around the hermeneutical relationship of faith, allegory and science. The first will lead us back to Augustine, as we will distinguish medieval allegory from allegory in the early church (§4). The second focuses on the literal reading of Genesis as expressing a mindset that is as much scientific as it is literary, for which we will turn to Thierry of Chartres and William of Conches (§5). The precarious balance between myth and science in Hugh of St. Victor will be the subject of some concluding remarks (§6).

4. *Augustine and Eriugena on enchanted creation*

a. *Augustine*

Proceeding from a Manichaean to a Platonic perspective, Augustine was preoccupied with Genesis for most of his life, as he faced the question how all things could come from a single, immaterial God. Through the exegesis of Ambrose, he adopted the view of an immaterial God. Even when mapping out his interior quest for the self in his *Confessions*, Augustine still informs us at length about his evolving views of creation. Surprisingly, he ends his convoluted conversion story with an extensive eulogy of the creator, as he clearly regards the quest for self-understanding and the understanding of creation as closely intertwined.

Written around the same time, Augustine's *On Christian Doctrine* develops essential guidelines for Christian exegesis. Recalibrating early Christian allegory, Augustine designs a revised method which is transported into the Middle Ages. David Dawson has summarized the twist of Augustinian allegory as follows:

[17] For a more nuanced view of both science and exegesis in the twelfth century, see P. Dronke, *Fabula: Explorations into the Uses of Myth in Medieval Platonism*, Leiden 1985.

> Ancient allegorical readings of Scripture have often been regarded as the means by which interpreters translated the unique images and stories of the Bible into the abstractions of classical metaphysics and ethics, but Augustine's recommendations concerning how to interpret Scripture suggest that nonliteral translation ought to move in the opposite direction. Rather than dissolving scriptural language into non-scriptural categories, allegorical reading should enable the Bible to refashion personal experience and cultural ideals by reformulating them in a distinctively biblical idiom.[18]

Developing the right view of creation might well be seen as one of the so-called cultural ideals mentioned by Dawson. In contrast to Ambrose, Augustine sees allegory as a way to increase biblical influence rather than restating it philosophically. His is not an allegory in the conventional Platonic sense, as the readers are invited to commit to the larger dynamic of the biblical text rather than accepting a specific interpretation. This strategy may explain also why we do not find the same focus on the creation of humanity in Augustine, with its gender division reflective of a psychological hierarchy of mind (*nous*) and senses (*aisthesis*).[19] As a result of Augustine's biblical expansionism, the creation story soon burgeons into a veritable meta-story, revealing and anchoring the being and belonging of God and the world.

Augustine's *De Genesi ad litteram* reflects this same holistic approach. Rather than containing a mere literal commentary, its structure involves a careful word-for-word explanation. In his profound respect for biblical idiom, Augustine considers no biblical words redundant. Reflecting the interdependence of cosmological and anthropological explanation in Genesis' ultra-natural theology, the work is divided into two main parts. Books I–IV concentrate on the six days of creation, after which the seventh day marks a day of rest. Books VI–XI deal with concrete creation, Adam and Eve, paradise and the first sin. Book V forms a transition and book XII deals with St. Paul's vision of paradise.[20]

---

[18] See D. Dawson, 'Sign Theory, Allegorical Reading, and the Motions of the Soul in *De doctrina christiana*', in: D. Arnold and P. Bright Kannengiesser, *De doctrina christiana—a Classic of Western Culture*, Notre Dame 1995, 123.

[19] Eriugena, who belonges in part to this older tradition, takes this Philonic interpretation from Ambrose's *De paradiso*. See Johannes Scottus Eriugena, *Periphyseon* IV 815C–D (ed. E. Jeauneau with a transl. by J.J. O'Meara, Dublin 1995), 174–175.

[20] See for the work's plan Augustine, *De Genesi ad litteram libri duodecim*, vol. 1, 18–25 (ed. and transl. by P. Agaësse and A. Solignac [Bibliothèque augustinienne 48], Paris 1972).

Augustine regards the creation of heaven and earth as the direct work of God and a true *creatio ex nihilo*. Taking great pains to give a meaningful reading of the opening sentence: 'In principio creavit Deus caelum et terram', his exegetical sensitivity carries his cosmology 'praeter allegoricam significationem' (I.1.2) back to Genesis as a literary rather than a literal text. Thus he asks the interesting question why God did not say: 'Let there be heaven and earth' in the same way as he said: 'Let there be light.' Apparently, the opening sentence carries special importance, as the totality of heaven and earth must contain all spiritual and corporeal reality. God's creation of heaven and earth 'in principio' signals that their immanence in the Word is to be distinguished from their material existence evoked by God's speaking ('Dixit Deus: fiat'). Only by turning towards God as the Word, clinging to this eternal Form, did creatures receive their own perfect form. Underlining the importance of the Trinity, Genesis points occasionally to the working of the Spirit, as it hovers above the waters, or informs the goodness of creation proclaimed by God ('uidit deus quia bonum est').[21]

Of key importance in the creation of heaven and earth are the *rationes aeternae*. These eternal reasons reveal how God's act of speech through certain intelligible locutions produces temporal creation.[22] More than the historical interpretation of paradise or of the creation of Adam in God's image, his view of *rationes aeternae* seals the connection between the eternal God, and the material world of growth and corruption existing in time. As a biblical exegete Augustine remained firmly interested in connecting divine transcendence with created immanence, even if it took him beyond conventional allegorical practice.

b. *Eriugena*

Augustine's *rationes aeternae* were adopted as *causae primordiales* by Johannes Scottus Eriugena in the ninth century. Eriugena's *Periphyseon* maintains the connection between the eternal Word and the divine attri-

---

21 The above paragraph reflects a summary of Augustine's arguments from *De Genesi ad litteram* I.I.2 to I.VI.12 (BA 48), 84–98.

22 See Augustine, *De Genesi ad litteram* I.IX.17, BA 48, 104: 'Sed multum est ac difficillimum capere, quomodo dicatur Deo non temporaliter iubente neque id temporaliter audiente creatura, quae contemplatione ueritatis omnia tempora excedit, sed intellectualiter sibimet inpressas ab incommutabili Dei sapientia rationes, tamquam intellegibiles locutiones, in ea, quae infra sunt, transmittente fieri temporales motus in rebus temporalibus uel formandis uel administrandis.'

butes, just as it upholds the difference between divine transcendence and the material world. Fusing Augustine's Forms in the divine mind with Dionysius' Divine Names or processions, however, Eriugena develops his own fourfold theory of nature as a Platonic synthesis of sorts.[23] *Natura* is divided into four forms, whose sequence reflects the cosmological pattern of *exitus* and *reditus*. These are (1) that which creates and is not created (*natura creans et non creata*) or God, (2) that which creates and is created (*natura creans et creata*) or the primordial causes, (3) that which is created and does not create (*natura non creans et creata*) or spatio-temporal creation, and (4) that which does not create and is not created (*natura non creans et non creata*) or God as final cause. Each form has a single book of the *Periphyseon* devoted to it, with the last form requiring two books.

With Eriugena's speculations closely linked to Genesis, some interpreters have regarded the *Periphyseon* as an extended *Hexaemeron*.[24] While that may be too sweeping a conclusion, Eriugena is keenly aware of the exegetical importance of rules. Halfway the third book he states that he will first engage in a literal interpretation of Genesis only thereafter to resort to allegory.[25] His literal interpretation, which follows Basil of Caesarea's *Homilies on the Hexaemeron*, continues until he embarks on his exegesis of the sixth day at the beginning of book IV. The introduction of allegory coincides with a remarkable transition in his story of Nature's forms, as he switches focus here from *exitus* to *reditus*. Overriding theme in Eriugena's exegesis of the sixth day is the paradise story, with its dramatic turn of events for humanity. In conformity with the Philonic line of interpretation found in Gregory of Nyssa and Ambrose, Eriugena considers the 'second' creation story of Gen 2 a mere elaboration of Gen 1:27b (man and woman he created them). Set apart from God's archetypal creation of humanity in his own image, humanity's physical creation and sexual division become explicitly linked to its fallen state.

By recounting the paradise story under the *aegis* of return, however, Eriugena's exegesis acquires an unexpected eschatological dimen-

---

[23] See Gersh, 'Cratylus mediaevalis', 85–88 on the fusion of the Augustinian-Boethian and the Dionysian tradition in Eriugena.

[24] So, famously, Guy-H. Allard, 'La structure littéraire de la composition du *De divisione naturae*', in: J.J. O'Meara and L. Bieler (eds), *The Mind of Eriugena*, Dublin 1973, 147–157.

[25] See Johannes Scottus Eriugena, *Periphyseon* III 693C (ed. I.P. Sheldon-Williams with a transl. by J.J. O'Meara, Dublin 1981), 196.

sion. As the reading of creation becomes now geared towards humanity's actual accomplishment of the return, the balance of cosmology and anthropology tilts towards the anthropological. Eriugena's unusual position that Adam's creation rather than his fall serves as the starting-point for cosmic return separates him also from the more linear cosmic views of Augustine.[26] Representing a unique brand of medieval idealism, Eriugena goes so far as to consider humanity's physical creation the material reflection of its superior and unified status in the divine mind as primordial cause.[27] As allegorical interpretation is ideally suited to bring out this deeper level, the *Periphyseon*'s theory of nature gravitates ultimately to the realms of metaphysics and theology.

5. *Twelfth-century thought about creation: changing dynamics of a Christian-Platonic world view*

a. *Introduction*

Moving into the twelfth century, we encounter the following situation. Due to the impact of Augustine's *On Christian Doctrine*, biblical exegesis had taken over in most intellectual endeavours, as early medieval theology was on the whole co-extensive with biblical interpretation. Augustine's overall exegetical influence notwithstanding, his *De Genesi ad litteram* was neither the sole nor the major source of information for speculation about creation. There seem to be two reasons for this. Firstly, the concept of creation was taking on a more dogmatic status as an important locus in the budding genre of early scholastic theology. Thought about creation now had to fit in with one's entire theological construct, the salvific purpose of which Abelard fittingly described as the summary knowledge of faith, love and the sacraments.[28] Secondly, the focus of studying Genesis was shifting. In the twelfth century

---

26 On the interrelatedness of procession and return, and the awkward coincidence of humanity's status *ante* and *post peccatum* to which it leads, see W. Otten, 'The Dialectic of the Return in Eriugena's *Periphyseon*', *Harvard Theological Review* 84 (1991) 399–421.

27 Eriugena's famous definition to this effect is found in *Periphyseon* IV 768B (ed. Jeauneau), 64: 'Possumus ergo hominem definire sic: Homo est notio quaedam intellectualis, in mente diuina aeternaliter facta.'

28 See Peter Abelard, *Theologia 'Scholarium'* I.1 (ed. Buytaert and Mews [CCCM 13], Turnhout 1987), 318: 'Tria sunt, ut arbitror, in quibus humanae salutis summa consistit, fides uidelicet, caritas et sacramenta.'

the primary purpose of reading Genesis was to test and exemplify ideas that one had developed otherwise, serving more as a meta-physical end-goal than a biblical starting-point.

Given the age's heightened interest in the sciences of the *quadrivium* alongside the *trivium*, moreover, the *Timaeus* played more into the sensitivity of its authors and scholars than Genesis. William of Conches, for example, was motivated by a strong desire to grasp nature's underlying structure as well as its organic harmony, even though his general aim still was to praise the creator by understanding creation. The definition he used described philosophy as 'the true comprehension of the things that are and are not seen and those that are and are seen'.[29] Reminiscent of Origen's approach in the *Peri Archon*, William associates true being, i.e., the true being of nature, specifically with the existence of things unseen, for which the myth of the *Timaeus* provided twelfth-century scholars with a rewarding source. While throughout the Middle Ages the proof text from Romans 1:20 ('Invisibilia enim ipsius, a creatura mundi, per ea quae facta sunt, intellecta, conspiciuntur') had legitimated the study of nature as an avenue to God, the surplus value of the *Timaeus* was that it allowed scholars to see God's invisible traits reflected in the arrangement of visible creation. Besides, what was there to fear from this pagan source? Even Plato respected the proper interdependence of cosmology and anthropology by putting requisite stress on salvation. As Abelard notes, the famous cross in *Timaeus* 36 B-C, the letter *chi* through which the World Soul keeps the motion of the cosmos under control, mystically points to the passion of the cross of the Lord (cf. also Van den Berg, this volume, §4).[30]

b. *Thierry of Chartres*

While Genesis and the *Timaeus* were more or less read in tandem, we notice how the latter work begins to encroach on the reading of Genesis in such works as Thierry of Chartres' *Tractatus de sex dierum operibus* (d. 1140s) and William's *Dragmaticon Philosophiae* (d. 1144–1149). Thierry's treatise offers an interpretation of Genesis *secundum phisicam et*

[29] See William of Conches, *Philosophia* I, I §4 (ed. Maurach, Pretoria 1980), 18: 'eorum quae sunt et non videntur, et eorum quae sunt et videntur vera comprehensio'. William derived this definition from the proëmium to Boethius' *De arithmetica*.

[30] See Abelard, *Theologia Christiana* II.16 (ed. Buytaert [CCCM 12], Turnhout 1969), 140.

ad litteram.[31] He wants to explain the causes from which the world drew its existence and the temporal order according to which they unfolded *secundum phisicam*. In the six days of creation four worldly causes were operative, namely an efficient cause (i.e., God), a formal cause (i.e., divine wisdom), a final cause (i.e., divine benignity) and a material cause (i.e., the four elements). Taking his cue from the first words of Scripture ('In principio creavit Deus celum et terra'), Thierry describes how God as efficient cause ('Deus') first created the material elements ('celum et terra'). The formal cause features in the Genesis-phrase 'And *God said*,' with God's act of speech referring to the ordering principle of divine wisdom. Divine benignity acts as the final cause, as seen in the recurrent phrase 'And *God saw that it was good*', with God's vision of creation mirroring his love.

Thierry's distinction between the efficient, formal and final cause allows him to integrate the Christian notion of the Trinity with his physical exegesis.[32] The Father represents creative power, the Son is that wisdom which brings order out of chaos, and the Spirit is divine benignity, as God created out of love. Seeing the four elements (fire, water, earth and air) jointly as material cause, Thierry mitigates the divide between creator and creation by intimating that the world somehow contributed to its own generation (see also Dillon's reply to Van Woudenberg, this volume). His repeated insistence that God created heaven and earth, i.e., fashioned the four elements, attempts to secure divine transcendence by salvaging the *creatio ex nihilo* idea.[33]

With the four elements now in place, creation can actually unfold from just these. By explaining creation not as dependent on a single divine act, but as a gradual development from secondary causes—

---

[31] See Thierry of Chartres, *Tractatus de sex dierum operibus* 1, in: N.M. Häring (ed.), *Commentaries on Boethius by Thierry of Chartres and His School*, Toronto 1971, 555: 'De septem diebus et sex operum distinctionibus primam Geneseos partem secundum phisicam et ad litteram ego expositurus, inprimis de intentione auctoris et de libri utilitate pauca premittam.' For a longer analysis of this section, see Otten, 'Nature and Scripture', 272–276.

[32] Thierry says so much in *Tractatus* 3.54–56 (ed. Häring), 556–557: 'Nam Pater est efficiens causa Filius uero formalis Spiritus sanctus finalis quatuor uero elementa materialis. Ex quibus quatuor causis uniuersa corporea substantia habet subsistere.'

[33] In *Tractatus* 3.50–54 (ed. Häring), 556, Thierry sees matter as a joint product of the Trinity: 'In materia igitur que est quatuor elementa operatur summa Trinitas ipsam materiam creando in hoc quod est efficiens causa: creatam informando et disponendo in eo quod est formalis causa: informatam et dispositam diligendo et gubernando in eo quod est finalis causa.'

an innovation probably introduced by Bernard of Chartres[34]—Thierry breaks rank with the tradition of Augustinian Genesis interpretations.[35] Just as the first day in Genesis brought on the next, the generation of one thing led naturally to another. On the first 'day', which Thierry calls the first 'integral revolution of heaven',[36] God created not only the four material elements, but also light, which originates from the highest element of fire. As fire warmed the highest part of the lower element, it illumined the air. Having generated light, the heat of fire next began to warm the third element, i.e., water. As the water surface evaporated and ascended into the air, it turned into clouds. Depicting God as putting the firmament 'firmly' in the middle of the waters on the second day, Genesis describes the physical process of air settling in between the evaporated waters and those below.[37]

Due to its author's innovative approach, Thierry's work poses as a scientific showpiece in the modern, anti-fideist sense. But the question is whether such a conclusion is warranted. Rather than rejecting allegorical interpretation, Thierry's literal approach may well reveal precisely his great respect for the authority of the Genesis text. Peter Dronke has argued that the *Tractatus* brings out how the physical 'unfolding' of the universe closely matches the *explicatio* of human speech in verbal language.[38] References to the laws of physics neither demythologize

---

[34] For Bernard's so-called use of the *formae nativae* as mediating between the absolute ideas and matter, see Bernard of Chartres, *Glosae super Platonem* (ed. by Paul E. Dutton, Toronto 1991), 70–96. Bernard may have been Thierry's older brother.

[35] See Augustine, *De Genesi ad litteram* I.XV.29 (BA 48), 120: 'Non quia informis materia formatis rebus tempore prior est, cum sit utrumque simul concreatum (Sir 18:1), ... formatam quippe creauit materiam ...' Augustine compares God's creation of formed matter with human speech. Just as one emits the sound and the specific pronouncement of a word simultaneously, so God created matter and form together. For the relative positions on creation of the Chartrians Thierry and William of Conches as well as of Hugh of St. Victor, see Charlotte Gross, 'Twelfth-Century Concepts of Time: Three Reinterpretations of Augustine's Doctrine of Creation *Simul*', *Journal of the History of Philosophy* 23 (1985) 325–338. Gross considers the Chartrian view of time as the duration of cosmic disposition an innovation, which she contrasts with Hugh's view of time as a sequence of historical events, ordained by God and effected for humanity's restoration (327).

[36] See Thierry, *Tractatus* 4. 58–59 (ed. Häring), 557: 'Dies naturalis est spacium in quo una celi integra conuersio ab ortu ad ortum perficitur.'

[37] See Thierry, *Tractatus* 8. 4–6 (ed. Häring), 558: 'Et tunc aer aptus fuit ut FIRMAMENTUM appellaretur quasi firme sustinens superiorem aquam et inferiorem continens: utramque ab altera intransgressibiliter determinans.'

[38] See P. Dronke, 'Thierry of Chartres', in: P. Dronke (ed.), *A History of Twelfth-Century Philosophy*, Cambridge 1988, 374.

Genesis nor disqualify the study of creation as a way to reach God. On the contrary, an essential aspect of Thierry's teaching is his view that knowledge of God is taught through God's works.

c. *William of Conches*

As they both depend on literal support, literary and scientific interests can actually go hand in hand. What Thierry managed to do for Genesis, William of Conches' *Glosae super Platonem* did for the *Timaeus*. As his *Glosae* explain in more detail, Plato's *Republic* had already dealt with positive justice, after which the *Timaeus* concentrated on natural justice, that is, on the creation of the world. Hence this was to be William's central theme also.[39]

William's preference for the gloss differs from Thierry's preference for literal commentary. Whereas a commentary collects the true meanings of the text (*in unum colligere*), the aim of glossing a text is to focus on the letter or continuation of the text (*continuatio litterae*) alongside its meaning, in a clear attempt not to separate them. In a gloss one must expound the words of the text in such a way as if the tongue of the doctor himself (*scil.* Plato) were uttering them.[40] For both of these masters, Plato's status was comparable to Vergil, Cicero, and Moses. When William faces the difficulty that Platonic philosophy harbours also some unchristian claims, he simply jokes that one cannot expect a pagan to do everything right.[41]

Two brief examples help to demonstrate the similarity of interest that lies behind William's and Thierry's physical interpretations. The first focuses on chaos, the second on the waters above the firmament. When William's *Dragmaticon Philosophiae*, written after the *Glosae*, speaks about creation, he comments on it as a joint product of created nature and a transcendent creator. William limits God's direct intervention to three feats: his creation of the elements and the human souls out of nothing,

[39] See William of Conches, *Glosae super Platonem* 3 (ed. Jeauneau, Paris 1965), 59.

[40] See *Glosae super Platonem* 10 (ed. Jeauneau), 67: 'Commentum enim, solam sententiam exsequens, de continuatione vel expositione litere nichil agit. Glosa vero omnia illa exequitur. Unde dicitur glosa id est lingua. Ita enim aperte debet exponere ac si lingua doctoris videatur docere.'

[41] William is aware of the discrepancies separating Plato from scriptural truth, but can use this disadvantage to the philosopher's favour. Thus he exclaims: 'Sed quid mirum si achademicus (i.e., Plato) alicubi achademice loquatur? Si enim ubique bene diceret, achademicus non esset.' See *Glosae super Platonem* 119 (ed. Jeauneau), 210–211.

Christ's birth from a virgin, and the resurrection of the dead. He held that God first created a big body, which the philosophers called chaos, as all the elements in it were mixed. With God next leaving the scene, we enter the realm of *opus naturae*, where nature holds sway. Nature always fashions something rude and mixed first, only thereafter to form and shape it. But why? The end of the *Dragmaticon*'s first book gives the following answer:

> Therefore, because nature and the craftsmen were unable to come up to the Creator's work, the Creator determined to come down to their standard. For, if this were not so, it would be thought to be a weakness in nature whenever things were created mixed by her. Or, as others say, God created mixed things to show how much confusion of things was possible if his own love were not ordering them.[42]

God's act of condescension to nature here is surprising and not easy to explain, especially since in his earlier chaos-theory William had deliberately refuted the idea that God ordered the elements in place to manifest his power. Obviously, God can do anything, even create a calf out of a tree trunk (cf. Tieleman, this volume, §3),[43] but the question is why he should choose to do so? Here it becomes manifest that it is not God's power, but his love depending on his will that brings order to creation. For William, the creator's love 'naturally' translates into the harmonious arrangement of creation. Radiating a beauty that is inclusive and near-divine, William conceptualizes creation increasingly in terms of harmony and symmetry. His purpose in doing so is to bring out better nature's own inherent principles, for *natura operans* ultimately accomplishes all this. For William, nature's arrangement is as much an aesthetic as a regulated physiological affair.

Sharing Thierry's literary sensitivity, William directs his scientific approach especially to the things unseen. The waters above the firmament form an interesting case in point. Far from arbitrarily assuming divine error, as when Peter Abelard laconically stated that even God

---

[42] See William of Conches, *Dragmaticon Philosophiae* I.7.4 (ed. Ronca [CCCM 152], Turnhout 1997), 31: 'Quia igitur natura et artifex non poterant ad operationem creatoris ascendere, uoluit creator ad illorum operationem condescendere. Si enim hoc non esset, debilitas naturae putaretur, quociens ab ea aliqua mixta crearentur. Vel, ut alii dicunt, mixtim creauit ut significaret quanta confusio rerum esse posset, nisi sua dilectio res ordinaret.' For the English translation, see I. Ronca and M. Curr (transl.), *William of Conches: A Dialogue on Natural Philosophy*, Notre Dame 1997, 18.

[43] See William of Conches, *Philosophia* II, II §5 (ed. Maurach), 43.

was at a loss about them,[44] William uses the invisibility of the supracelestial waters to inspect the laws of physical causality more closely. Rejecting Bede's solution that these waters were frozen, as the firmament would have collapsed under the weight of the ice, his alternative solution is to see them as evaporated waters, that is, as plain air.[45] Following the *Timaeus* rather than Genesis, William pursues his own explanations and analogies by concentrating increasingly on the intrinsic aesthetics of nature's *modus operandi*. Favouring the exploration of *natura operans*, he regards this intermediate level between God's prime act of creation from nothing (*opus creatoris*) and the ordinary handiwork of human beings (*opus artificis*) as the near-perfect expression of God's ordering love in cosmological disguise.[46]

6. *Conclusion: the precarious balance of myth and science*

Both Thierry of Chartres and William of Conches initially came to their material as teachers of the liberal arts. As such, they stood in a tradition of expounding the texts of a respected master, be it Moses or Plato. In line with the exegetical turn brought about by Augustine, who himself had been thoroughly trained in the liberal arts, they aimed at designing their own 'ultra-natural' theology by integrating the physics of Genesis with the metaphysics of the *Timaeus*. To conclude from their respective methods, i.e., the literal approach and the gloss, that they should be seen as forerunners of a modern scientific outlook belies how they nowhere depart from the method of textual interpretation. While their rejection of conventional allegory may echo Augustine's approach to Genesis, their literary sensitivity reveals them to be close also to the

[44] This is why God did not say on the second day that he saw that it was good. See Abelard, *Expositio in Hexaemeron* (PL 178), 740A.

[45] See *Philosophia* II, I–II (ed. Maurach), 41–44. William's discussion of the waters above the firmament derives from his larger discussion of the four elements, as the supracelestial waters are located in the region of the upper air or aether, which consists of fire. He rejects the literal interpretation of Gen 1:7 that God separated the waters above the firmament from those below the firmament as *contra rationem*, see *Philosophia* II, I §3. See also H. Rodnite Lemay, 'Science and Theology at Chartres: the Case of the Supracelestial Waters', *The British Journal for the History of Science* 10 (1977) 226–236.

[46] For a fuller analysis of William's philosophy, see W. Otten, 'Plato and the Fabulous Cosmology of William of Conches', in: J. Spruyt and M. Kardaun (eds), *The Winged Chariot: Collected Essays on Plato and Platonism in Honour of L.M. de Rijk*, Leiden 2000, 185–203.

programme of the known poets of the age. Bernard Silvestris and Alan of Lille likewise tried to recast the universe's genesis as an imaginative process involving philosophical and theological recreation.

Rather than drawing them into heresy, it seems that the unique combination of biblical physics and Platonic metaphysics allowed Thierry and William to draw parallels, notice analogies and make new connections in their own creative way. In this respect their approach is in full conformity with the practice of *integumentum*, a rhetorical strategy whereby one seizes on a text's poetic and philosophical polyvalence to pry it open and unveil a kernel of underlying Christian truth, rather than foreshadowing thirteenth-century natural science.[47]

Such a conclusion about the vitality and creativity of this method inevitably brings up the question why this kind of associative thinking ultimately went out of style. In my view this is primarily due to the growing disconnection between exegesis and the liberal arts, causing the Platonizing thought of the twelfth century to disintegrate and substituting it with the more orthodox and standardized approach of scholasticism. Without deep roots in rhetorical practice, the role of allegory changes from intellectual habit to formulaic exegesis, as allegory lays the foundation for a mystical super-structure. As an example we may look to Hugh of St. Victor's theological summa *De sacramentis*, which introduces a sharp division between the so-called *opus creationis* and the *opus restaurationis*. Pagan writings may still be read to find out about creation, but only Scripture discloses the work of restoration. While Hugh reveals himself in other respects to be a master of the literal sense, the authority of the Bible differs radically from all other *auctoritates* for him, which is not unlike how theology distances itself from philosophy and the arts. With the main purpose of the letter of the Bible seen as underlying a higher, spiritual reading, the goal of exegesis is now to instruct us about the mysteries of the faith (*sacramenta fidei*), with knowledge of creation taking second place.[48]

---

[47] See for an analysis of William of Conches and Thierry of Chartres in this latter sense A. Speer, *Die entdeckte Natur: Untersuchungen zu Begründungsversuchen einer 'scientia naturalis' im 12. Jahrhundert*, Leiden 1995, 130–288.

[48] See Hugh of St. Victor, *De sacramentis*, prol. cap. 2 (PL 176), 183A–B: 'Materia diuinarum Scripturarum omnium, sunt opera restaurationis humanae. Duo enim sunt opera in quibus uniuersa continentur quae facta sunt. Primum est opus conditionis. Secundum est opus restaurationis … Ergo opus conditionis est creatio mundi cum omnibus elementis suis. Opus restaurationis est incarnatio Verbi cum omnibus sacramentis suis; siue iis quae praecesserunt ab initio saeculi, siue iis quae subsequuntur usque ad finem mundi.'

Once theology and the liberal arts went their separate ways, the divergence between scientific interpretation and biblical exegesis was a by-product that inevitably followed. Predating this divergence, Augustine had studied Genesis for both reasons, as for him the world's beginning (*initium mundi*) was inherently related to its beauty (*exornatio*), giving him ample reason to praise the creator. Taking the *initium*-side more or less for granted while concentrating increasingly on the *Timaeus*, twelfth-century scientists favoured explaining the *exornatio mundi*.[49] Although it never was the intent of these scientists to confine the reading of Genesis to the *initium*-side, from the acceptance of a basic crack in the canon of twelfth-century classics other problems were bound to arise. Thus Thierry's position that the spontaneous transition from inanimate to animate nature does not require the intervention of a creator at all (see also Dillon's reply to Van Woudenberg, this volume),[50] or the idea that human bodies are formed from the four elements could easily give rise to charges of heresy. On the latter point William interestingly attributes the weaker state of woman to the fact that she is made from less-balanced clay.[51]

With the ongoing development of science along Aristotelian lines in the thirteenth century, not only do we see how the tension between science and exegesis results in separation, but also how the isolation of theology becomes a fact, as it severed not just its intrinsic ties with science, but especially its organic ties with myth.

[49] See B. Stock, *Myth and Science in the Twelfth Century: A Study of Bernard Silvester*, Princeton 1972, 277. As Stock argues convincingly, although the point cannot be pursued here, the idea of *exornatio* is closely tied to the morality of the cosmos.

[50] See P. Dronke, 'Thierry of Chartres', 375, who argues that Thierry goes further even than William of Conches.

[51] This is William's non-literal interpretation of Eve's creation from Adam's rib. See his *Philosophia* I, XIII §42–43 (ed. Maurach), 37–38.

DESIGN IN NATURE: SOME CURRENT ISSUES

René van Woudenberg

1. *Design in nature and Darwinism*

Far into the eighteenth century many scientists and philosophers, if not most of the ordinary folk, believed that the physical-biological world shows signs of being created by God. The eighteenth-century Scottish philosopher and scientist Thomas Reid, for instance, was a Newtonian and involved in applying the inductive method to various areas of scientific research. At the same time he believed that the world displays signs of design.[1] In his *Essays on the Intellectual Powers of Man*, he formulates a principle to the effect 'that design and intelligence in the cause may be inferred, with certainty, from the marks or signs of it in the effect'.[2] Although the principle does not specify 'marks or signs of design and intelligence', from scattered passages in his work we may conclude that what Reid had in mind are such features as contrivance, order, organization, intent, purpose, usefulness, adaptation, aptness of means to ends, regularity, and beauty. So, from the presence of these features in objects, so the principle tells us, it may be inferred that those objects are designed, and hence that a designer of those objects exists. According to Reid not only artifacts like hammers, watches, and houses etc. display the indicated features; he held also 'that there are in fact the clearest marks of design and wisdom in the works of Nature',[3] one example of which is 'the structure of the human body'.[4] Reid's principle, then, licenses the inference that whatever displays such marks, is designed and brought forth by an intelligent cause, either human or divine.

---

[1] For an assessment of Reid as a scientist, see Paul Wood, 'Thomas Reid and the Culture of Science', in: Terence Cuneo & René van Woudenberg (eds), *The Cambridge Companion to Thomas Reid*, Cambridge, 2004, 53–76.

[2] Thomas Reid, *Essays on the Intellectual Powers of Man* (ed. D. Brookes), Edinburgh 2002, 504.

[3] Reid, *Intellectual Powers*, 509.

[4] Reid, *Intellectual Powers*, 510.

Expounding on the principle, Reid makes it clear that he thinks that an object's displaying the marks cannot be due to chance. With approval he quotes Cicero who had said, in *On Divination* (1.23):

> Can any thing done by chance have all the marks of design? Four dice may by chance turn up four aces; but do you think that four hundred dice, thrown by chance, will turn up four hundred aces? Colours thrown upon canvas without design may have some similitude to a human face; but do you think they might make as beautiful a picture as that of the Coan Venus? A hog turning up the ground with his nose may make something of the form of the letter A; but do you think that a hog might describe on the ground the Andromache of Ennius?[5]

So Cicero's (and Reid's) point is that certain features cannot be ascribed to chance but only to design—thus signalling their belief that design and chance are mutually exclusive (see also Tieleman, this volume, §2).

Now Reid, of course, lived in the pre-Darwin era. Darwin's theory of evolution, however, rendered application of Reid's principle to the biological world obsolete, hence 'arguments from design' without force and belief in design without justification. At least, so many have claimed. This claim rested on the observation that many of the marks mentioned above, especially adaptation (to local environment) and aptness of means to ends, at least in so far as they seem to occur in the world of living animals, can be accounted for by means of Darwin's theory, i.e. by means of a theory that makes no reference whatsoever to design. In succinct form, Darwin's theory boils down to the following set of theses:

1. All organisms tend to produce more offspring than can possibly survive.
2. Offspring vary among themselves.[6]
3. At least some of this variation is passed down by inheritance to future generations.
4. If [1], i.e. if not all offspring can possibly survive, and if [2], i.e. if offspring vary among themselves, then, on average, survivors will tend to be those individuals that happen to be best suited to the local environment. Since [3] is true, i.e. since hereditary exists, the offspring of survivors will tend to resemble their successful parents.

[5] Cited from Reid, *Intellectual Powers*, 505.

[6] Darwin spoke of 'descent with modification'. We should keep in mind that Darwin did not know the mechanism of heredity (Mendel's principles) nor did he specify the source of the variation [2] talks about (later Darwinists have specified that mechanism as random genetic mutation).

The accumulation of these favourable variants through time will produce evolutionary change.

Thesis [4] is generally referred to as 'the principle of natural selection'. By means of it Darwinists explained the adaptedness of organisms to their environments without referring to design. The existing organisms display adaptedness, they said, only because ill-adapted forms have been eliminated by natural selection. Appeal to design or a divine designer seems superfluous.

Darwin's theory constituted a death-blow for all theorizing about the living world in terms of design. Since Darwin, arguments from design (arguments for the conclusion that there is a God, from premises about observed design in the world) are therefore highly suspect. In recent decades, however, ideas about design seem to make something of a come-back. The task of this chapter is to show where and how that is the case.

2. *Darwinism and design: are they incompatible?*

As indicated, it has often been repeated that Darwinism meant the death stroke for the thesis that living creatures are products of divine design. One reason for thinking that Darwinism had undermined that thesis was that according to Darwinism chance plays an uneliminable role in the emergence of the various species of living creatures. And the general idea was that where there is chance, there can be no design, and where there is design there can be no chance. A number of philosophers, however, have recently argued that Darwinism and design are compatible. This is, of course, an important point, for if Darwinism and design are compatible, then the truth of Darwinism cannot, all by itself, refute the thesis that the various living species are products of design.

In order to be able to see how the arguments for the compatibility of Darwinism and design go, we need to be clear on what we take Darwinism to be, for Darwinism is different things to different people. It seems to me that we need to distinguish a least the following theses:

1. Progress Thesis: there was unicellular life before there was multicellular life, there were worms before fishes, fishes before amphibia, amphibia before reptiles, birds, and mammals; and finally there are human beings.

2. Common Ancestry Thesis: life originated on one place on earth; all living creatures are literally distant cousins of one another.
3. The Thesis of Natural Selection: the mechanism that is responsible for the progress of simple to complex forms of life, as well as for the formation of the various species, is natural selection working on some source of randomness, e.g. random genetic mutation.
4. Naturalistic Origins Thesis: life itself developed from non-living matter without any divine creative activity but just by virtue of the ordinary laws of physics and chemistry.

It should be noted that these theses are logically independent of one another, except for 3, which presupposes 2 (it would not make sense to propose a mechanism for the emergence of more complex forms of life, if one does not think the evolution has indeed occurred). For example, the Progress Thesis does not entail the Common Ancestry Thesis, nor does the latter entail the former. Again, the Thesis of Natural Selection does not entail the Naturalistic Origins Thesis, nor is it entailed by it. Let us now turn to the question what we shall take 'Darwinism' to be the name of? Is it the name of the set of all four theses? Or is it a subset of them? Since different authors take different approaches here, I will simply announce that I will use 'Darwinism' to denote the Thesis of Natural Selection.

In a famous quotation, where he speaks of events that have been identified as the sources of mutation, Jacques Monod has hinted at the incompatibility of Darwinism and design:

> We call these events accidental; we say that they are random occurrences. And since they constitute the only possible source of modifications in the genetic text, itself the sole repository of the organism's hereditary structure, it necessarily follows that chance alone is at the source of every innovation, of all creation in the biosphere. Pure chance, absolutely free but blind, at the very root of the stupendous edifice of evolution: this central concept of modern biology … is today the sole conceivable hypothesis, the only one that squares with observed and tested fact.[7]

Mutations, the only source of biological innovation, says Monod, are random occurrences, and hence all biodiversity is the result of 'free but blind chance'. This last expression signals Monod's denial of the thesis that the biodiversity is the product of intelligent design.

[7] Jacques Monod, *Chance and Necessity: An Essay on the Natural Philosophy of Modern Biology*, New York 1971, 112–113.

Contrary to what Monod claims, however, Darwinism, i.e. the Thesis of Natural Selection, is not incompatible with the thesis that the living species display design. As Peter van Inwagen has argued, to suppose that they are incompatible, is to commit the fallacy of composition, the fallacy one commits when one reasons 'because a cow is entirely composed of quarks and electrons, and quarks and electrons are non-living and invisible, a cow must therefore be non-living and invisible'.[8] The argument for this claim is that it is possible that God brought about the enormous diversity of living species through natural selection. The idea is that it is possible that God, so to speak, 'used' natural selection 'as a means', or an instrument, to attain his purposes, and that this possibility is not forestalled by the fact that this instrument involves an ineliminable element of chance.

In order to make this clear, Van Inwagen draws an analogy with a device for calculating the areas surrounded by irregular closed curves that works according to what is sometimes called the dartboard technique. Suppose you draw such a curve on a screen; then the device randomly selects points on the screen, and then looks at each point to see whether it falls inside or outside the curve, as the number of points chosen increases, the ratio of the chosen points that fall inside the curve to the total number of points chosen tends to the ratio of the area enclosed by the curve to the area of the screen. Every point on the screen is randomly selected, but the device in which the randomizer is built-in serves a certain goal that is not due to chance: the goal of calculating areas surrounded by irregular closed curves. Likewise, even if Monod is right and every mutation that ever occurred is due to chance, it does not follow that every aspect of the biosphere is due to chance. Even if none of the mutations has a purpose, it does not follow that the biosphere has no purpose. To think otherwise is, to say it again, to commit the fallacy of composition. Therefore, a process that involves ineliminable chance is capable of being used as a means to attain a certain goal.

The same point can also be made via a slightly different route. Mutations, we are told by Monod and others, are chance events. But what, exactly, does this mean? Any textbook on evolutionary biology will give you the following answer: mutations are chance events in the sense that mutations do not occur in response to changes in the

---

[8] Peter van Inwagen, 'The Compatibility of Darwinism and Design', in: Neil A. Manson, *God and Design: The Teleological Argument and Modern Science*, London 2003, 354–355.

environmental perils or opportunities: there is no correlation between the 'usefulness' of a particular mutation and the likelihood that it will occur. The thesis that mutations are due to chance in the sense indicated, however, is compatible with the thesis that God has been guiding evolution by deliberately causing certain mutations. Mutations are not responses to environmental perils and opportunities, but this by no means rules out the possibility that God brought them about.

There is, however, another sense in which mutations may be called chance events and that is such that one might think that if mutations are chancy in that sense, any process that involves chance in that sense is incapable of being used as an instrument by God or any other agent. What I have in mind is this: something may be called 'due to chance' when it is uncaused, as is the case with certain quantum events. Monod thinks that mutations are 'due to chance' also in this sense: they are uncaused.[9] And one may think that something that involves uncaused events cannot be anyone's 'means' to attain a certain goal.

But the thesis that mutations are due to chance in the sense of being uncaused is, as Del Ratzsch has argued, compatible with the thesis that certain aspects of the biosphere are products of design.[10] His argument supposes that there are true counterfactuals of chance. A counterfactual is a statement that says that something would be the case if something else had been the case. 'If John had been offered a bribe, he would have refused it' is a counterfactual—a counterfactual of freedom (thus named because the statement tells us what a certain person would do out of his own free will). One might suppose that counterfactuals have truth values, i.e. that it is either true or false that John would have refused the bribe, had he been offered one. One might furthermore suppose that just as there are counterfactuals of freedom, there are counterfactuals of chance. The basic idea is this. If we think that the basic laws of nature are indeterministic in character, then there would be no way for God to create some radioactive atom that would be causally guaranteed to decay exactly at a given time. Yet it is possible that it would decay at precisely that moment. Suppose now it *would* decay at that moment. Then God would know that fact. Thus, although nothing would *cause* decay at that moment, the following counterfactual of chance is nonetheless true: 'Were God to

[9] 'A mutation is in itself a … quantum event', *Chance and Necessity*, 114–115.

[10] Del Ratzsch, 'Design, Chance, and Theistic Evolution', in: William Dembski (ed.), *Mere Creation*, Downer's Grove 1998, 303–306.

create an atom in such and such a condition, it would in fact decay at that precise moment'. One might suppose that God, knowing the truth values of such counterfactuals, reckons with them in order to bring about whatever he intends to bring about. If so, the thesis that mutations are uncaused is compatible with the thesis that God designed the biosphere.

Many philosophers, however, reject the idea that counterfactuals of freedom have truth value, and those who do will almost certainly also reject the idea that counterfactuals of chance have truth value. Let us, if only for the sake of the argument, suppose that counterfactuals of chance don't have truth value. In that case God could not know in advance when a certain quantum event is to take place and hence he cannot reckon with it when he plans to bring something or other about. Would this not imply that God could not use anything that involves this kind of chance as a means to bring about whatever he intends to bring about? Would this not imply the incompatibility of the thesis that mutations are chance events (in the sense of being uncaused events) and the thesis that God created the biosphere? As Ratzsch has argued, it does not. His argument consists in the depiction of a possible scenario he calls 'subjunctive supervision'. In this scenario God initiates the process that eventually leads to the coming about of the biosphere and God intends the coming about of the biosphere. It is assumed that God does not know when any of the mutations necessary for the coming about of the variety of species will occur. Still, God oversees the evolutionary process and is prepared to act whenever chance threatens to turn the process in a wrong direction. As a matter of fact, however, the process stays on track and God never has to intervene in the chance processes. If this scenario were to obtain, it would still make good sense to say that God designed the biosphere and that he brought it about by means of a process that involved chancy, i.e. uncaused, events.[11]

To round off this discussion, I would like to make a few remarks. First of all, those who argue for the compatibility of Darwinism and design need not be committed to the truth of either Darwinism or design, nor to the truth of both of them. The compatibility claim is

[11] Further arguments for the compatibility of Darwinism and Design can be found in René van Woudenberg, 'Darwinian and Teleological Explanations: Are They Incompatible?', in: Philip Clayton and Jeffrey Schloss (eds), *Evolution and Ethics: Human Morality in Biological and Religious Perspective*, Grand Rapids 2004, 171–186; René van Woudenberg, *Toeval en ontwerp in de wereld*, Budel 2003, 13–50.

a claim as to what is, or is not, excluded by Darwinism or design and this claim can be made irrespective of any commitment as to the truth value of Darwinism or design.

Secondly, the previous point implies that the arguments reviewed in this section by no means establish that the biosphere is in fact designed. So, in the context of a discussion about the truth or falsehood of Darwinism, compatibility arguments have no force whatsoever. The only context in which such arguments are relevant is a discussion in which it is claimed that because mutations are due to chance, the biosphere is not designed. In other words, compatibility claims are only relevant in a context in which design claims are countered by an appeal to chance events in the evolutionary process.

Thirdly, compatibility claims are typically made by philosophers—they typically are the products of armchair reflection, i.e. the kind of reflection that is required for conceptual analysis (such as the analysis of such concepts as 'chance' and 'design') as well as for logical analysis (that has to do with implicative relationships between statements); compatibility claims are not the result of empirical investigation. The next section is devoted to a highly interesting empirical claim concerning design, one that also directly relates to the truth value of Darwinism.

3. *Darwinism and irreducibly complex systems*

In the *Origin of Species* Darwin proposed an explanation of the great variety and complexity of the biosphere and, as indicated in the previous section, his explanation referred to the mechanism of natural selection working on some source of random variation. Darwin did not propose his explanation as being apriori correct, nor as necessarily true.[12] As a matter of fact, he indicated that certain empirical data, if they were to turn up, would constitute problems for his theory: 'If it could be demonstrated that any complex organ existed which could not possibly have been formed by numerous, successive, slight modifications, my theory would absolutely break down'.[13] Darwin reported that he

[12] At least one of Darwin's followers, however, has suggested that Darwinism is an apriori truth. Richard Dawkins, 'Darwinism Triumphant: Darwinism as a Universal Truth', in: Michael Robinson and Lionel Tiger (eds), *Man and Beast Revised*, Smithsonian Institution Press 1991, 23–39.

[13] Charles Darwin, *On the Origin of Species*, New York 1999 (1958), 154.

could find no such organs and he placed the burden of proof, i.e. the proof that there exist complex organs that could not possibly have been formed by numerous, successive, slight modifications, on his opponents. Nevertheless, Darwin formulated a criterion such that if something satisfies it, it would constitute a counterexample to and hence a problem for his theory. As he says, if such a counterexample were to be found, his theory would absolutely break down. Now, what might satisfy Darwin's criterion? The biochemist Michael Behe has argued that 'irreducibly complex systems (or organs)' satisfy Darwin's criterion for being a successful counterexample. In his book *Darwin's Black Box: The Biochemical Challenge to Evolution* he defined an irreducible complex system as 'a single system which is composed of several well-matched, interacting parts that contribute to the basic function, and where the removal of any one of the parts causes the system to effectively cease functioning'.[14]

As an example of something that is irreducibly complex in this sense, Behe refers to a mousetrap. This mice-catching device has several parts, including a wooden platform, a spring with extended ends, a hammer, holding bar and catch. If one of these parts is missing, the trap would not catch any mice: *all* of the parts have to be in place at the same time, if it is to catch mice. Just a few of the parts simply would not do the job. The removal of any one of the parts causes the mousetrap to cease functioning. Therefore it is an irreducibly complex system.

Behe's main contention is that as biology has progressed with dazzling speed in the past half-century, many systems in the cell have been discovered which, like a mousetrap, are irreducibly complex. He supports this claim by giving detailed and technical descriptions of various such systems. Let me quote Behe's own summary description of the bacterial flagellum:

> The flagellum is quite literally an outboard motor that some bacteria use to swim. It is a rotary device that, like a boat's motor, turns a propeller to push against liquid, moving the bacterium forward in the process. It consists of a number of parts, including a long tail that acts as a propeller, the hook region that attaches the propeller to the drive shaft, the motor that uses a flow of acid from the outside of the bacterium to the inside to power the turning, a stator that keeps the structure stationary in the plane of the membrane while the propeller turns, and bushing material to allow the drive shaft to poke up through the bacterial membrane. In the absence of the hook, or the motor, or the propeller, or the drive shaft

[14] Michael Behe, *Darwin's Black Box: The Biochemical Challenge to Evolution*, New York 1996, 39.

> or most of the forty different types of proteins that genetic studies have shown to be necessary for the activity or construction of the flagellum, one doesn't get a flagellum that spins half as fast as it used to, or a quarter as fast. Either the flagellum doesn't work, or it doesn't even get constructed in the cell. Like a mousetrap, the flagellum is irreducibly complex.[15]

So far, this is only description. But this description serves as the basis for the point that it is very difficult to envision how the flagellum could have developed by 'numerous, successive, slight modifications'. To see the difficulty, just try to envision a trail through space-time such that it begins with a very simple system that, through chance events that are responsible for very small phenotypic changes all of which confer evolutionary advantages to the emerging systems, ends with something as complex as the flagellum. Behe's point with respect to irreducibly complex systems such as the flagellum, then, is that it is very hard, if not impossible, to see how they can be explained by invoking natural selection that operates on random mutation. The reason for this is that the stages that supposedly precede the stage where the entire complex system is in place, are such that they could not have possibly survived the selective pressures—which Darwinian theory tells they must, for the only changes that it allows are very small ones.[16]

Behe does not just claim that it is hard, or impossible, to envision an evolutionary scenario leading up to, for instance, the flagellum. He also claims that nowhere in the scientific literature a serious and detailed model for how the flagellum might have arisen in a Darwinian manner has ever been proposed, let alone that experiments have been conducted to test such a model. Therefore, the flagellum seems to meet the criterion of being a serious counterexample to Darwin's theory.

The flagellum, as well as other irreducibly complex systems, therefore require an alternative explanation. Behe's own proposal is that such systems be explained by 'intelligent design'. This proposal contains a positive and a negative part. The negative part is that such systems are *not* designed by the laws of nature, nor by chance and

---

[15] Michael Behe, 'The Modern Intelligent Design Hypothesis', in: Neil A. Manson (ed.), *God and Design: The Teleological Argument and Modern Science*, London 2003, 280.

[16] Sometimes Behe seems to suggest that irreducibly complex systems defy Darwinian explanation *by definition*, as when he says 'An irreducibly complex system cannot be produced directly by slight, successive modification of a precursor system, since any precursor to an irreducibly complex system is by definition non-functional', 'Molecular Machines', in: Robert Pennock (ed.), *Intelligent Design Creationism and its Critics*, 247.

necessity, i.e. not by non-intelligent agents. The positive part is that such systems are *planned* by a designer with intelligence who knew what the systems would look like when they were completed. In order to forestall misunderstanding of Behe's position, I should like to emphasize that in *Darwin's Black Box* Behe argues that the flagellum, as well as the other examples he gives of irreducible complex systems, exhibit design regardless of whether they were produced by natural (albeit non-Darwinian) means or by supernatural agent intervention at some point in the past.

Not being a biochemist myself, I am not properly positioned to evaluate Behe's ideas and proposals. But I do think that I am in a position to make some cautionary remarks. First of all, Behe's case for the claim that irreducible complex systems have not been and cannot be explained in Darwinian fashion is a straightforward scientific claim that nowhere rests on religious premises. Behe's claim is a claim to the effect that there are certain phenomena that a certain theory cannot explain, and this is certainly not a sort of claim that is foreign to scientific debates. Behe's claim is, of course, controversial but this cannot be a reason to discard it out of hand. After all, many new scientific ideas were once controversial. Furthermore, Behe has not just thrown out unripe ideas but he has carefully made a case for his claim. His claim is founded on recent findings in biochemistry concerning the molecular basis of life—a claim that is up for discussion. I should add that a number of critics have tried to refute Behe's claim. But also that Behe responded to many if not most of them in a manner that gives both bystanders and experts in the field the impression that Behe is on to something.[17]

Secondly, Behe's proposed alternative explanation in terms of intelligent design should also be taken as a proposal *within the bounds of science*. And like all scientific theories, there may eventually be very good reasons to discard it. Not everybody, however, will grant that 'intelligent design' is a concept that can legitimately be used in a scientific theory. As a matter of fact, many intellectuals take it for granted that 'intelligent design' cannot be used in scientific theorizing. It is impossible to discuss this matter in any depth. Two remarks must suffice. (i) In everyday contexts we habitually engage in design explanations all the time. 'Why is that book on my desk?' I ask; and the explanation may

[17] For an overview of criticisms I can do no better than refer to Behe's webpage, where they are all listed. That page also contains a number of responses to his critics.

be 'because Jane wants you to read it'. 'Why isn't John at the meeting?', you may ask; I explain it is because John wants to sabotage the negotiations. But if we engage in design explanations all the time, why should we stop doing that once we enter the halls of science? (ii) There is no such thing as 'the essence of science' that somehow rules that design explanations are always out of order. In fact, science is an evolving practice and concepts and ideas that once seemed illegitimate may later become acceptable.[18]

Thirdly, Behe's argument is an argument for the conclusion that irreducibly complex systems are designed; it is not an argument for the conclusion that a benevolent God exists. Behe argues for design, but the question as to the identity of the designer is left open. There are various possible candidates to fit that role, e.g. the God that Christians believe in, angels, Plato's demiurge, some mystical force, space aliens, or some utterly unknown intelligent being. But Behe's scientific argument is not committed to any of these—for it is compatible with the existence of any of them. This feature of Behe's argument sets it apart from the famous design arguments of William Paley who argued for the conclusion that design is due to a benevolent God.[19]

Finally, Behe claims that intelligent design is a good explanation for a number of biochemical systems. He emphatically does not propose intelligent design as an explanation for everything. As a matter of fact, Behe acknowledges that evolutionary biologists have recognized a number of factors that might have affected the development of life: common descent, natural selection, migration, population size, founder effects, genetic drift, gene flow, linkage, and more. The claim that some biochemical systems were designed by an intelligent agent does not mean that any of the other factors are not operative, common, or important.[20]

---

[18] That science has no essential nature I have argued in my book *Toeval en ontwerp in de wereld*, 51ff. A most helpful discussion of arguments for the impermissibility of design explanations in science is Del Ratzsch, *Nature, Design, and Science: The Status of Design in Natural Science*, New York 2001, 105–147.

[19] It should be noted that some authors who offer design arguments are unlike Behe in that they hold that such arguments do not commit one to the existence of a designer. E.g. Ronald Meester, 'Ontwerp in natuurwetenschap: een zinvol concept?', *Radix* 30 (2004), 1–20.

[20] Behe, 'Molecular Machines: Experimental Support for the Design Inference', in: Robert T. Pennock (ed.), *Intelligent Design Creationism and its Critics*, Cambridge, MA 2001, 255.

4. *A fine-tuned cosmos*

Another area of discussion where design ideas have been newly introduced is cosmology. For a long time such ideas were virtually absent in that field of study. So what happened? The main thing is that the scientific view of the cosmos has drastically changed over the last few decades. Ever since the nineteenth century the cosmos was considered to be an amorphous affair consisting of matter in motion. Recent cosmology, however, has shown that our cosmos hangs in a delicate balance, that it is 'fine tuned' so as to be a friendly abode for life. The idea is that our cosmos could easily have been radically different from the way it actually is, and that if it had only been slightly different, there would have been no life. These facts suggest, so it is argued, that the fine tuning of the cosmos is due to design. In this section I will give an impression of various ideas that go into this line of thinking.

In the last twenty years a number of books have appeared in which it is argued that even slight changes in any of the so-called 'cosmological constants' would produce a dramatically different cosmos, one that is unsuitable for life of any conceivable type. Some of the more important and influential ones are John Barrow and Frank Tipler's *The Anthropic Cosmological Principle*,[21] John Leslie's *Universes*,[22] Paul Davies' *The Accidental Universe*,[23] Paul Davies' *The Cosmic Blueprint*,[24] and Gilles Cohen-Tannoudji's *Universal Constants in Physics*.[25] Among the cosmological constants are the so-called 'Universal Constants': Boltzmann's constant, Planck's constant, the Speed of Light, and the Gravitational Constant. Among the cosmological constants are furthermore the so-called 'Fine Structure Constants': the Gravitation fine structure constant, the Fine structure constant of weak interaction, the Electromagnetic fine structure constant, and the Fine structure constant of strong interaction. The cosmological constants finally include the masses of the elementary particles, such as the mass of the proton, electron, and neutron and the unit charge for the electron or proton. All the cosmological constants have a certain value, for instance, the value of the Gravitational Constant $G = 6.67 \times 10^{-11}$ N-m^2/kg^2. If the cosmos is to be

21 Oxford, 1988.
22 New York, 1989.
23 Cambridge, 1982.
24 Portsmouth, 1988.
25 New York, 1993.

life-permitting, so these authors argue, there needs to be a very precise balancing of the nuclear strong force and the electromagnetic force. This balancing is required if there is to be the abundance of carbon in nature that is needed for the development of life. So there is this 'cosmic coincidence' that the values of the strong force and the electromagnetic force are such that carbon-based life is possible. And these authors argue that there are many more of such coincidences.

To facilitate thinking about these matters (at least for a non-physicist) it may be helpful to think of the cosmos as the product of a machine designed to produce cosmoi.[26] The machine has, let us say, some twenty or thirty dials on it. The overall features of the cosmos are the result of the ways the dials were set when the cosmos was produced. Had the dials been set in other positions, a different type of cosmos would have emerged from the machine. The lesson that the authors I have mentioned teach is that many statements of the following form are true:

- the pointer on dial 18 is set at 0.0054321; if it had not been set at some value between 0.0054320 and 0.0054322, there would be no carbon atoms, and hence no life.
- the pointer on dial 23 is set at 5.113445 and the pointer on dial 5 is set at 5.113449; if the values of these dials had been exactly equal, there would have been no matter, but only radiation; if the two readings had differed by more than 0.000006, all stars would be of a type that would burn out before multicellular organisms could evolve on their planets.

When it is said that our cosmos is fine-tuned, what is meant is twofold. Firstly, that the dials on the machine that produced our life-containing cosmos were set between quite strict parameters. And secondly, that only a vanishingly small proportion of the ways the dials could be set would make the machine produce a cosmos that is life-permitting.

This fanciful way of thinking inevitably leads up to the question 'Why were the dials set right for life?', 'Why is our cosmos one of the few possible cosmoi that is a friendly abode to life?' In response to this question, a number of people have framed an answer in terms of design. The dials were set right for life, they say, because some rational agent designed it that way.

---

[26] This way of putting the issue is due to Peter van Inwagen, *Metaphysics*, Boulder 1993, 130.

This answer, of course, has met with considerable resistance, nor is it the only possible answer. An alternative, and most popular, answer is that it is sheer accident, pure luck that the dials were set right for life. After all, so the reasoning goes, if the machine has dials, the dials have to be set *some* way, and any particular way is as likely (or better, since the probabilities are very small, as *un*likely) as any other; hence, there is nothing particularly remarkable about the setting that resulted in the current cosmos, nothing that needs what I call 'a special explanation'. For, to say it again, given that the dials have to be set in *some* way, the setting for the current cosmos is as unlikely as any other setting.

As proponents of the design explanation of the fine-tuning have pointed out, however, this reasoning should convince nobody. What is wrong with it can be brought out by reference to what Leslie has called 'the principle of the merchant's thumb'. There is a story behind the principle. It is about a merchant who displays an expensive silk robe to a potential buyer. The robe has a hole in it, but the merchant consistently keeps it covered with his left thumb. When he is accused of dishonestly concealing the hole from the buyer, he and his defenders point out that everyone's left thumb has to be *somewhere* and that the merchant's left thumb being right above the hole is as likely (or as unlikely) as its being at any other place. There is therefore, they say, nothing remarkable about its being above the hole, it does not need any special explanation. But, of course, no one should accept this piece of reasoning. What the story suggests is that there is a principle to the following effect:

> Principle of the Merchant's Thumb: if we can think of a 'special' explanation of a certain fact that would be a very good explanation, then it is wrong to discard that explanation simply because no such special explanations are available for otherwise similar facts.

The chance explanation of the settings of the machine's dials violates this principle and should therefore not be accepted.

Yet another explanation of the settings of the dials is that, somehow, these settings are *the only possible* settings, that, contrary to what was suggested, the various cosmological constants could not possibly have other values. Those values would then, somehow, be necessary. In response to this, however, the proponents of design have pointed out that at the moment there is no reason at all to suppose that the values *are* necessary.

Given what was just said, some proponents of design reason as follows: 'If the setting of cosmic dials cannot be explained by reference to (natural) necessity, nor by reference to chance, then, by elimination, only one possibility is left, viz. that the setting is intentionally designed by a rational agent'. Those who reason thus,[27] may, but need not, make use of William Dembski's 'explanatory filter' that tells us that if we want to explain some phenomenon, we should first of all see if it can be explained in the sense that it fits a general pattern; if it cannot be thus explained, we should conclude that the phenomenon is due to chance, unless the probability that the phenomenon occurs is so small, that it must be explained by reference to design.[28]

Two comments will suffice to round off this section. Firstly, the argument advanced by the proponents of design has been countered by the 'multiple worlds' hypothesis, the hypothesis that there are, or have been, very many cosmoi. According to this hypothesis the cosmic machine has produced not just one cosmos but a real plethora of them. On this hypothesis, it is vastly less improbable that our cosmos is a friendly abode to life than on the hypothesis that there is, and has been, only one. This is so for the very same reason as that it is less probable that one will throw snake eyes in one throw, than it is that one will throw snake eyes in one hundred throws. But if the many cosmoi hypothesis is true, then 'chance' is a very good explanation of our cosmos' being life-permitting; *that* explanation does not violate the Principle of the Merchant's Thumb. So, if the many cosmoi hypothesis is true, then there is no rationale for a design inference.

The cogency of this argument depends, of course, on the likelihood of the hypothesis. Some think it is a rather plausible hypothesis, others find it badly *ad hoc*, advanced only so as to be able to escape the otherwise inescapable design inference. This is not the place to enter into this discussion, but it seems clear that discussion is needed here.

Secondly, the proponents of design, as I have depicted them, argued for a design explanation of the setting of the cosmic dials by means of an elimination argument. Since this setting cannot be explained by reference to some (natural) regularity, nor by chance, what is left is the design explanation. This way of arguing for design, however, is rather problematic. For, as was argued in section two, design and chance do

---

[27] William Lane Craig, 'Design and the Anthropic Fine-Tuning of the Universe', in: Neil A. Manson (ed.), *God and Design*, London 2003, 161–175.

[28] William Dembski, *The Design Inference*, Cambridge 1998, 36–66.

not necessarily exclude one another. Furthermore a rather strong case could be made for the claim that (natural) regularity and design do not necessarily exclude one another. But if that is true, then any argument for design that works by eliminating both regularity and chance, will give unwanted results!

The task of this paper was to indicate where and how ideas about design have recently resurfaced in discussion about nature. They have resurged in philosophy, biochemistry and in current cosmology. Philosophers have argued that claims to the effect that something is designed cannot be countered by showing that the coming about of that something involves chance. Michael Behe argued that irreducibly complex biochemical systems defy Darwinian explanations and that the best alternative explanation is in terms of design. Finally, a number of cosmologists have argued that our cosmos is fine-tuned and that this requires an explanation in terms of design.

It would be a gross overstatement to say that ideas about design in nature have become acceptable to the average scientist. Many of them are hostile to the very suggestion. But that hostility, it seems to me, stems not so much from scientific findings as it does from preconceived opinions about what science is, and about the impermissibility of design explanations. These opinions are, most of the time at least, philosophical opinions—and about such opinions, I fear, a true consensus is not available. Or, to put the same point in a more positive vein, about such opinions a genuine, albeit not irrational, dissensus is possible.

DESIGN IN NATURE: SOME COMMENTS FROM THE ANCIENT PERSPECTIVE

John Dillon

I found René van Woudenberg's paper most stimulating when I heard it, and I have been very glad to have had a chance to read it over subsequently. At the time, I recall commenting that what is at issue between Darwin and such modern followers of his as Richard Dawkins, on the one hand, and a critic such as Michael Behe on the other, resembles to some extent—though on a very much more sophisticated plane—the ancient conflict between an atomist such as Democritus and a partisan of purposive divine creation such as Plato.

George van Kooten has asked me to expand on this for the publication of the proceedings of our conference, and I am glad to do so, even if, perforce, only briefly and superficially. What I had in mind was this. In the period of Plato's youth, during the last quarter of the fifth century BCE, the philosophers Leucippus and Democritus had developed a radical theory of cosmology, involving the theory that the whole complexity of the world as we know it is made up of nothing more than combinations of 'atoms'—minimum units of matter—in motion in an infinite void. There is an infinite number of atoms—somewhat confusingly, of a number of different shapes and sizes[1]—moving at random in this void, and, over an infinite extent of time, they hook up with each other in a great many different combinations, ultimately to produce the great variety of entities, and levels of entity—that is to say, not just inanimate objects, but living things, and conscious intelligent subjects—that we observe to exist. In our world there is no purpose to any of this, though ineluctable laws of nature do arise later, to govern the actions of the various materials formed, once they have been formed. Even human consciousness and free will is to be explained—though the explanation, as we have it, is the later contribution of the philosopher Epicurus, who took up the atomist theory a century after Democritus, at the end of the fourth century—by the postulation of

[1] The reason for this, in Democritus' mind, was that one had somehow to explain the different degrees of consistency and solidity of existent things, but it is a postulate that generates at least as many problems as it purports to solve.

the existence of certain very fine, smooth atoms, and of certain unpredictable 'swerves' or jumps that they take (rather like the postulates of quantum theory, perhaps).

All this was vehemently opposed by the philosopher Plato. So repugnant is Democritus to him that he never condescends to mention him by name in his works, but he is the enemy lurking in the background of Plato's own great essay at cosmology in the *Timaeus*, and of his attack on atheism in Book X of his last work, *The Laws* (cf. also Tieleman, this volume, §3, on Galen's criticism of Epicurus). For Plato it is axiomatic that something so well-formed as the cosmos could only be the purposive creation of a benevolent and provident divine creator, albeit working with an irreducibly recalcitrant substratum, or 'matter', which prevents him from generating a perfect product. For Plato, the physical world is constructed on the basis of an intelligible pattern, or 'paradigm', by the projection onto the chaotic material 'receptacle' of a co-ordinated set of formal principles, manifesting themselves as a series of combinations of basic triangles (see also Dillon's paper in the present volume). These triangles and their combinations (forming the five 'Platonic bodies', pyramid, cube, octahedron, dodecahedron, and icosahedron) constitute Plato's 'answer' to Democritus' postulate of atoms.

It is Plato's firm conviction that the random jostling of atoms, by itself, can never issue in an ordered universe. He does not attempt to prove this assumption, but he does go out of his way, in one notable passage of the *Timaeus* (52E–53B), to counter the Democritean position, even at the cost of appearing somewhat illogical himself (see also Dillon's paper, esp. §1). He has chosen to present his account of the Demiurge's creation of the world as a temporal process, which involves the postulation of a pre-cosmic chaotic state of the 'receptacle'. His immediate successors, Speusippus and Xenocrates, stoutly maintained that this was not intended to be taken literally, and I am inclined to agree with them, but it does present Plato with some interesting problems of exposition. In this case, he chooses to postulate that, before the Demiurge has imposed order on the Receptacle, certain 'traces' of the elementary triangles are already present in it, and cause it to be subject to disorderly motion. This motion is portrayed as being rather like the action of a winnowing-fan, which would tend to send like particles in the direction of like, and separate off heavy and light elements in opposite directions. *But*, says Plato, *this tendency to separation and distinction will never come to anything, without the intervention of the Demiurge.* It looks very much as if Plato is choosing to take on Democritus directly, especially as we happen to know, from a passage of Sextus Empiricus (*Adversus mathematicos* VII.117 = Diels-Kranz 68 B 164), that Democritus himself used the image of a sieve (κόσκινον)—not too different from a winnowing-fan (πλόκανον; *Timaeus* 52E)—to describe the separating-out of different types of atoms.

Plato is actually guilty of falling into illogicality here, it seems to me, in his concern to counter the Democritean world view, since he is unable to explain why, once the winnowing motion in the Receptacle has *begun* to separate out the 'traces' of the basic corpuscles, this process should not continue to a conclusion; he simply declares that, without the imposition of order by the Demiurge, it will never come to anything.

At any rate, that is the basic opposition between the purposive and non-purposive views of the cosmogonic process from antiquity.

One particular item in Prof. Van Woudenberg's exposition, however, attracted my special attention, and that was his account of Behe's theory of 'irreducibly complex systems' as a counter-example to Darwin's theory. Van Woudenberg expresses some doubts as to the validity of Behe's position (though modestly admitting his lack of expertise in biochemistry), and I would be inclined to agree with him—though very modestly having to admit that I am no scientist of any kind! However, it does not seem to me fatal to Darwinian, or indeed Democritean, theory that one should have to postulate, at certain levels of complexity in the development of a system, a sort of 'quantum leap', where, let us say, a certain concatenation of molecules *suddenly* produces a living organism (see also Otten, this volume, §§5b and 6) and another level of complexity, equally suddenly, produces an organism manifesting consciousness. This, if I understand the term rightly, is none other than the phenomenon of *emergence*, and I do not see why it cannot be supposed to happen quite randomly and automatically. The same would go for the emergence of a planetary environment which can sustain life—though one would logically have to allow here (as indeed does Democritus) for an infinite multiplicity of such worlds!

Here it may be that a consideration of the world-view of a thinker like Plotinus would be of some relevance. Plotinus' first principle, the One, is an entirely impersonal entity (though Plotinus would regard it as impersonally *benign*, and tending to the bringing about of an orderly world). But the One's generation of the cosmos is entirely automatic; it can do no other than it does, though Plotinus would regard what it does as being for the best; its creative activity is conditioned by the fact that a whole range of phenomena, including living things and rational intelligences, are required for the world to be a *world at all*—that is, a totality of all possible levels and varieties of being. This cosmogony is certainly not random, but the One cannot be said to *plan* it: it is simply the inevitable result of its own perfection, which spills over into creativity. So then, it seems to me, one can have a rationally ordered universe, a *design*, which is yet not planned by a personal creator. This might be seen as a third possibility, between the Darwinian and the Deist models.

INDEX OF ANCIENT TEXTS

I. *Ancient Near Eastern Texts*

II. *Greek and Latin Pagan Texts*

III. *Jewish Texts*

III.1. Hebrew Bible

III.2. Apocrypha and Septuagint

III.3. Pseudepigrapha and other Jewish Texts

III.4. Dead Sea Scrolls and Related Texts

III.5. Philo of Alexandria

III.6. Josephus

III.7. Targumim

III.8. Rabbinic Texts

IV. *Christian Texts*

IV.1. New Testament

IV.2. Early Christian and Medieval Writers

THEMES IN BIBLICAL NARRATIVE JEWISH AND CHRISTIAN TRADITIONS

ISSN 1388-3909

1. GARCÍA MARTÍNEZ F. & G.P. LUTTIKHUIZEN (eds). *Interpretations of the Flood.* 1999. ISBN 90 04 11253 7
2. LUTTIKHUIZEN, G.P. (ed.). *Paradise Interpreted.* Representations of Biblical Paradise in Judaism and Christianity. 1999. ISBN 90 04 11331 2
3. LUTTIKHUIZEN, G.P. (ed.). *The Creation of Man and Woman.* Interpretations of the Biblical Narratives in Jewish and Christian traditions. 2000. ISBN 90 04 11671 0
4. NOORT, E. & E. TIGCHELAAR. (eds). *The Sacrifice of Isaac.* The Aqedah (Genesis 22) and its Interpretations. 2002. ISBN 90 04 12434 9
5. LUTTIKHUIZEN, G.P. (ed.). *Eve's Children.* The Biblical Stories Retold and Interpreted in Jewish and Christian Traditions. 2003. ISBN 90 04 12615 5
6. AUFFARTH, C. & L.T. STUCKENBRUCK (eds). *The Fall of the Angels.* 2004. ISBN 90 04 12668 6
7. NOORT, E. & E. TIGCHELAAR (eds). *Sodom's Sin.* Genesis 18-19 and its Interpretations. 2004. ISBN 90 04 14048 4
8. VAN KOOTEN, G.H. (ed.). *The Creation of Heaven and Earth.* Re-interpretations of Genesis I in the Context of Judaism, Ancient Philosophy, Christianity, and Modern Physics. 2005. ISBN 90 04 14235 5